SCALING AND INTEGRATION OF HIGH-SPEED ELECTRONICS AND OPTOMECHANICAL SYSTEMS

SELECTED TOPICS IN ELECTRONICS AND SYSTEMS

Editor-in-Chief: **M. S. Shur**

---

*The complete list of the published volumes in the series can be found at
http://www.worldscientific.com/series/stes

Selected Topics in Electronics and Systems – Vol. 59

SCALING AND INTEGRATION OF HIGH-SPEED ELECTRONICS AND OPTOMECHANICAL SYSTEMS

Editors

Magnus Willander
Linköping University, Sweden

Håkan Pettersson
Halmstad University, Sweden

World Scientific

NEW JERSEY · LONDON · SINGAPORE · BEIJING · SHANGHAI · HONG KONG · TAIPEI · CHENNAI

Published by

World Scientific Publishing Co. Pte. Ltd.

5 Toh Tuck Link, Singapore 596224

USA office: 27 Warren Street, Suite 401-402, Hackensack, NJ 07601

UK office: 57 Shelton Street, Covent Garden, London WC2H 9HE

British Library Cataloguing-in-Publication Data

A catalogue record for this book is available from the British Library.

Selected Topics in Electronics and Systems — Vol. 59
SCALING AND INTEGRATION OF HIGH SPEED ELECTRONICS AND
OPTOMECHANICAL SYSTEMS

Copyright © 2017 by World Scientific Publishing Co. Pte. Ltd.

ISBN 978-981-3225-39-8

Printed in Singapore

Preface

What has been called the third revolution is under way; Manufacturing is going digital, driven by the computing revolution, and in particular, by CMOS technology and its development. Three manuscripts in this special issue illustrate aspects of this including scaling challenges for CMOS from an industrial point of view, SiGe BiCMOS technologies, circuits for mm-wave and THz applications, and process technologies for the 14 nm node and beyond. Two examples of niche applications for emerging technologies then follow: The first deals with carbon electronics and plasmonics for the THz region, while the second reviews nano-optomechanical systems on silicon chips.

- The first paper is from GLOBALFOUNDRIES in the US and written by Dr A. P. Jacob *et al*. The authors give a review of the latest innovations regarding devices and integration challenges in the CMOS industry.
- The second paper, from IHP in Germany is written by Dr A. Mai *et al*. and deals with high-speed (up to the THz-region) SiGe BiCMOS technologies.
- The third paper is from the Institute of Microelectronics in China. It is written by Dr H. Radamson *et al*., and reports on the problems associated with source/drain integration in FinFET technology nodes at 14 nm and beyond.
- The first of the two niche technology papers is written by Dr M. Ryzhii *et al*. from University of Aizu in Japan. It describes a carbon nanotube network and plasmonics for the THz region.
- The last paper, written by Zhu Diao *et al*. from Halmstad University in Sweden, reviews recent developments in integrated on-chip nano-optomechanical systems.

Special Issue on Scaling and Integration of High Speed Electronics and Optomechanical Systems.

Guest Editors

Prof. Magnus Willander
Linköping University, Sweden

Prof. Håkan Pettersson
Halmstad University, Sweden

Contents

Scaling Challenges for Advanced CMOS Devices

Ajey P. Jacob[*], Ruilong Xie, Min Gyu Sung, Lars Liebmann, Rinus T. P. Lee and Bill Taylor

GLOBALFOUNDRIES, 400 Stonebreak Road ext., Malta, New York 12020, USA
[*]*ajey.jacob@globalfoundries.com*

The economic health of the semiconductor industry requires substantial scaling of chip power, performance, and area with every new technology node that is ramped into manufacturing in two year intervals. With no direct physical link to any particular design dimensions, industry wide the technology node names are chosen to reflect the roughly 70% scaling of linear dimensions necessary to enable the doubling of transistor density predicted by Moore's law and typically progress as 22nm, 14nm, 10nm, 7nm, 5nm, 3nm etc. At the time of this writing, the most advanced technology node in volume manufacturing is the 14nm node with the 7nm node in advanced development and 5nm in early exploration. The technology challenges to reach thus far have not been trivial. This review addresses the past innovation in response to the device challenges and discusses in-depth the integration challenges associated with the sub-22nm non-planar finFET technologies that are either in advanced technology development or in manufacturing. It discusses the integration challenges in patterning for both the front-end-of-line and back-end-of-line elements in the CMOS transistor. In addition, this article also gives a brief review of integrating an alternate channel material into the finFET technology, as well as next generation device architectures such as nanowire and vertical FETs. Lastly, it also discusses challenges dictated by the need to interconnect the ever-increasing density of transistors.

Keywords: CMOS; fin; finFET; FEOL; BEOL; Bulk silicon; SOI; FDSOI; Nanowire FET; Vertical FET; design technology co-optimization (DTCO); self-aligned double patterning (SADP); self-aligned quadruple patterning (SAQP); channel engineering; gate engineering; source drain (S/D)engineering; contact engineering; fin pitch; contacted poly pitch or gate pitch (CPP); Epitaxial (epi); Silicon; Silicon Germanium (SiGe); Germanium; III-V; Indium Gallium Arsenide (InGaAs); source drain epi; replacement metal gate (RMG); Self-aligned contact (SAC); single and double diffusion; interconnect resistance; interconnect capacitance; interconnect patterning, technology node; 22nm; 14nm; 10nm; 7nm; 5nm.

1. Introduction

The end of semiconductor scaling, not just due to insurmountable technical challenges but also due to a saturation in the demand for more compute power, has been predicted and disproven repeatedly. After the transition from mainframe servers to desktop PCs, hand-held and smaller devices have now replaced bulkier laptops and mobile devices as the main consumers of advanced transistor technology. As these devices increase their presence in the internet of things (IOT) they demand higher data rate transmission, data storage, retention and faster data access from cloud databases. This means, that the demand for much smaller, faster and ultra-low power scaled devices continues to grow, and semiconductor scaling will continue for another decade or more[1].

The progression of semiconductor scaling is marked by 'technology nodes' that are refreshed on roughly a two year cadence. The expectation for every new technology node is to improve the power, performance and area (PPA) from the prior node. PPA scaling is driven by continuous improvements in transistor density as described by Moore's law and is expected to improve circuit performance by 30%, decrease the power consumption by 50% and reduce chip area by 50% with only minor cost-per-wafer increase from the previous node and no degradation in reliability[2]. Clearly, these aggressive scaling targets are difficult to maintain over many scaling cycles with new challenges arising at every technology node. In older technology nodes, scaling challenges at the device level, also referred to as 'front end of line' (FEOL), were primarily associated with short channel effects, i.e. the inability to completely turn off transistors at shorter gate lengths, while interconnect delay due to RC coupling was the primary concern in the wiring levels, also referred to as the 'back end of line' (BEOL). In addition to these electrical 'device and interconnect' challenges, advanced technology nodes have to contend with steadily increasing lithographic patterning barriers arising from the growing gap between available and required optical resolution. To maintain profitable scaling of PPA in spite of these challenges, several innovations have been implemented in the (1) complementary metal oxide semiconductor (CMOS) transistors (2) interconnects, and (3) sub-resolution patterning.

The first part of this paper (sections 2 and 3) will review the scaling challenges and associated innovations in the lagging CMOS nodes. This will prepare the reader for the second (section 4) and third (section 5 and 6) part of the paper which will discuss potential challenges and innovation opportunities for leading-edge CMOS nodes that utilize non-planar devices such as fin channel field effect transistors (finFETs). Innovation in leading-edge technology nodes involves advances in design technology co-optimization (DTCO), covered in section 4, materials and integration challenges in both FEOL and BEOL, discussed in section 5 and 6 respectively.

2. The Challenges and Approaches for CMOS Transistor Scaling

Innovations on the transistor for the lagging and leading technology nodes are depicted in Fig. 1.

The driving force for the innovations within the roadmap (Fig. 1) can primarily be explained by the need to maintain drive current in a metal oxide semiconductor (MOS) transistor, designated as I_{on} in Equation 1[3].

$$I_{ON} \sim \frac{1}{L} W C_{ox} \mu V_{dd}^{2} \tag{1}$$

where L is the channel Length, W is the width of the transistor, C_{ox} is the gate capacitance, μ is the surface mobility or effective mobility of the electrons or holes and V_{dd} is the source to drain voltage. The rest of the discussion in section 2 will correlate the above equation with the innovations depicted in the roadmap (Fig. 1) to explain challenges associated with channel and gate engineering (section 2.1), source drain (S/D) and contact engineering (section 2.2).

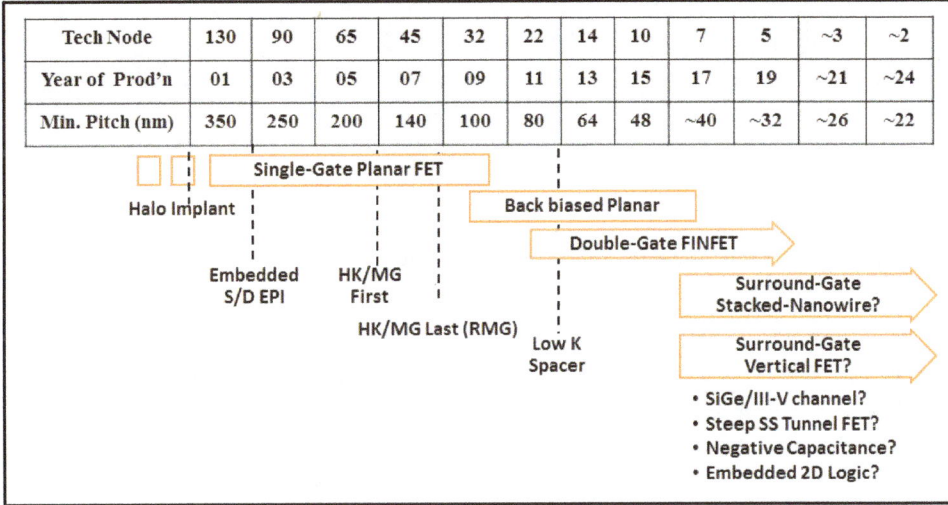

Tech Node	130	90	65	45	32	22	14	10	7	5	~3	~2
Year of Prod'n	01	03	05	07	09	11	13	15	17	19	~21	~24
Min. Pitch (nm)	350	250	200	140	100	80	64	48	~40	~32	~26	~22

Single-Gate Planar FET

Halo Implant

Back biased Planar

Double-Gate FINFET

Embedded S/D EPI

HK/MG First

HK/MG Last (RMG)

Low K Spacer

Surround-Gate Stacked-Nanowire?

Surround-Gate Vertical FET?

- SiGe/III-V channel?
- Steep SS Tunnel FET?
- Negative Capacitance?
- Embedded 2D Logic?

Fig. 1. FEOL Device Roadmap depicting two decades of key innovations.

2.1. *Channel and gate engineering*

The fundamental requirement in designing a nanoscale transistor is that, above all, it must hold together electrostatically. This simple principle is quintessential to understanding the past two decades of device evolution. The dominant element for 'holding together electrostatically' is the transistor gate which serves two key functions. In the 'ON' state, with maximum circuit voltage, V_{dd}, applied, the gate must attract enough minority carriers to provide a thick enough channel to allow sufficient current to flow. But the more important state for understanding device behavior is the 'OFF' state with no voltage applied, where the gate must exert a dominating electric field in the underlying silicon to strongly push away minority carriers and prevent drain induced barrier lowering that generates undesired source to drain leakage current[4]. This 'electrostatic dominance' implies overwhelming the influence of the source and drain regions which the gate cannot easily control. This is particularly important for the drain at V_{dd} which is creating a large depletion region under the gate. If the drain is deep, the lateral depletion region is also deep and thus regions farther away from the gate are difficult to control. Fig. 2 shows how simple channel length scaling leads to loss of gate control, but a shallow extension of the source and drain enables recovery of gate control. These shallow extensions are implemented through shallow implants followed by the formation of spacers to laterally offset the deep source and drain implants.

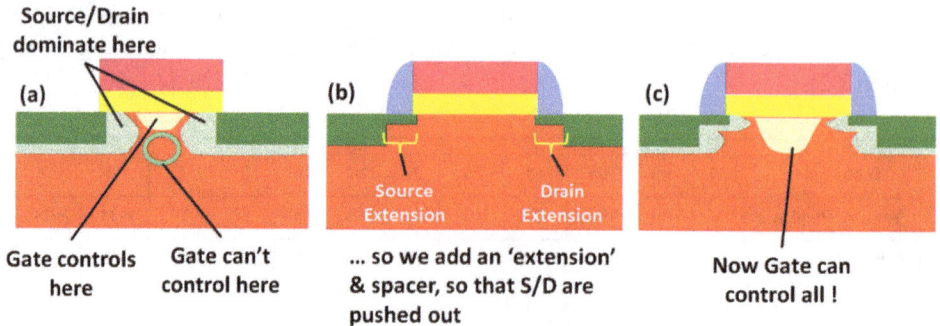

Fig. 2. "L" Gate scaling in a transistor (a) long channel transistor (b) short channel transistor with source/drain extension implants (c) short channel transistor depicting gate control with gate bias.

In addition to changing the structure of the source and drain regions, the OFF-state leakage can be reduced by increasing the doping in the well region. However this comes at a cost to ON current because carriers moving through the channel are inhibited by the presence of well dopants. An acceptable balance is obtained by keeping the middle of the channel lowly doped and locally increasing the doping under gate edges. This variable doping profile enables leakage resistance where it is needed at the source and drain, while maintaining good carrier mobility and low resistance in the center of the gate. This more complex doping profile is obtained through a tilted 'halo' implant, so-called because it creates a ring of doping around the gate. These two concepts, 'shallow extension' and 'local halo', have been fundamental to device scaling since the late 1980's.

Further scaling has required additional structural changes to the device, but every one of these changes can still be tied to gate control of the channel in the OFF state. One particular structural innovation for better gate control is the move from bulk silicon substrates to silicon-on-insulator (SOI) substrates to avoid other short channel effects such as punch through leakage current[5,6,7,8]. IBM made the first 64 bit power microprocessor in 0.22um CMOS SOI technology in 1997[9]. Simply put, SOI helps to control the gate by eliminating the deeper silicon bulk region. The gate's dominating electric field in the OFF state can deplete the minority carriers from the underlying silicon. For Si thicknesses of 20-40nm, some un-depleted silicon remains in what is referred to as partially depleted SOI (PDSOI) while for Si thicknesses of <10nm fully depleted SOI (FDSOI) is achieved. If the underlying (buried) oxide is thin enough, as in FDSOI wafers, further influence on both OFF and ON state behavior can be exerted by applying an electrical bias to the underlying bulk silicon to generate capacitive coupling to the top device Si. While such 'active biasing'[10] provides enormous potential for improved functionality on a chip, it is also responsible for the current shift in the way designers think about integrated circuits[11]. The biggest challenge with SOI wafers in manufacturing is its added cost. SOI wafers are fabricated through a process technique called 'Smart Cut', where a thin layer of bulk silicon is exfoliated (i.e. cleaved) after hydrogen ion implantation and annealing. The exfoliated silicon is bonded to a carrier wafer to form the SOI wafer. Fig. 3 compares MOS devices built on bulk silicon, PDSOI and FDSOI substrates. The image differentiates how the junction leakage can be reduced by using SOI structures. Researchers have been extensively studying more cost effective alternatives to SOI wafers by creating dielectrically isolated wafers through various integration schemes. Non-planar transistors such as fin channel field effect transistor (finFET) are better suited to dielectrically isolated technology since it is easier to achieve dielectric isolation through thermal processes in the fins than in planar transistors[12]. However, finFET technology carries its own manufacturing challenges.

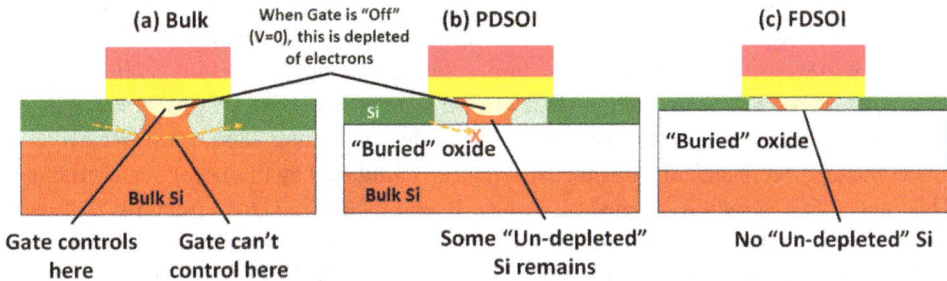

Fig. 3. Comparative image of devices built on (a) Bulk silicon (b) partially depleted SOI (PDSOI) and (c) fully depleted SOI (FDSOI) wafers. The sequence of images from bulk to PDSOI to FDSOI shows leakage reduction from bulk wafers to fully depleted SOI wafers.

It should be noted that the scaling of planar bulk transistors ended with the 20nm technology node because, as described above, the gate electrostatics could not hold together. Enhancement of the electrostatics in finFET comes by enabling two juxtaposed gates to control the channel[13]. While this double-gated design is not feasible to build in a planar structure (see Fig. 4), it is obtainable with non-planar manufacturing techniques in a finFET structure[14,15,16,17,18]. Currently manufactured finFETs have three gates, where the small portion running over the top of the fin is considered the third gate. By moving to this tri-gate device one can obtain the necessary gate control over the channel.

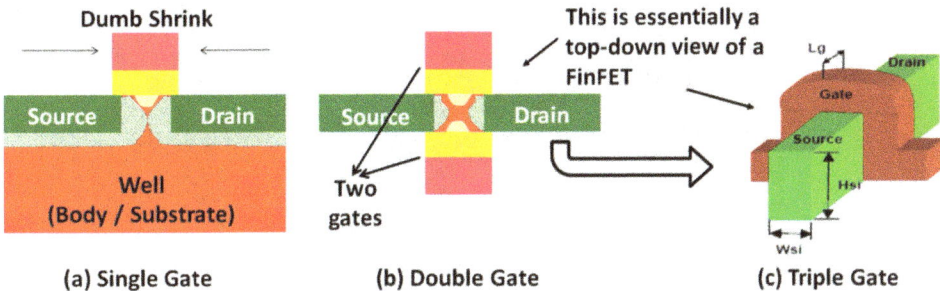

Fig. 4. Comparative images of (a) single gate planar structure with a dumb shrink (b) double and (c) triple gate devices; L_g, H_{si} and W_{si} corresponds to gate length, fin height and fin width respectively.

In addition to the electrostatic benefits, the three dimensional nature of finFETs, where the gate channels are formed on the vertical sidewalls of the fin and not just on the horizontal surface, is enabling more current flow per planar footprint. This satisfies the designers' need to pack more transistors per unit area without sacrificing device performance. However, as gate lengths shrink further beyond the 7nm technology node, even tri-gate finFETs have trouble maintaining gate control, leading to the need for gate all around (GAA) device structures also referred to as surround gate device structure. Impressive advancements in semiconductor process integration over the past 10 years have paved the way for GAA lateral nanowire and vertical transistors (see Fig. 5). The now-complete encirclement of the channel leads to enough additional gate control to maintain the needed leakage current at the yet-smaller gate lengths. However, it is clear that if one moves to a nanowire architecture, a single wire will be insufficient to meet the drive current requirements, and a stacked 2-or-3 high nanowire configuration will be necessary to enable the necessary gate control while obtaining enough drive current to overcome the "overhead" of parasitic capacitances introduced by this new device structure. This means that vertically aligned multi-gate lateral nanowire GAA structures still have substantial AC concerns. The vertical GAA FET structures can also be thought of as multigated vertically aligned structures (only NFET, only PFET or complimentary N & P vertically aligned FET) that improve density scaling but add additional complexities. In addition to structural or geometric innovations, it is to be noted that in silicon based systems, the gate lengths have been controlled/fabricated down to 3nm[19] which means gate length scaling can progress well beyond the ~14nm gate lengths of today's finFET. In addition, recent literature shows that gate lengths down to 1nm have been fabricated on alternate channel materials such as a molybdenum disulfide (MoS_2) transition metal dichalcogenide layered structures[20]. Excellent ON/OFF ratios up to ~10^6 have been obtained in these devices. Such transition metal dichalcogenides exhibit excellent microelectronics properties that have the potential to replace channel materials. These novel materials could be integrated onto the BEOL of the CMOS transistors as embedded logic devices, thus providing a mix of transistors in the circuit.

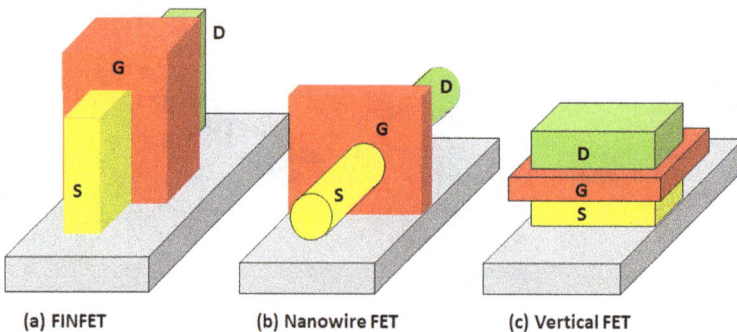

(a) FINFET   (b) Nanowire FET   (c) Vertical FET

Fig. 5. Potential Alternate Architecture for current and future technologies; (a) FinFET, (b) Nanowire FET (c) Vertical FET.

Another revolutionary enhancement in performance and reduction in gate leakage happened with the introduction of high-k metal gate devices at the 45nm technology node. The motivation for this innovation was to reduce the short channel effects, such as poly depletion and gate tunneling current, associated with polysilicon gates and ultrathin gate oxides respectively. Polysilicon has been used as the gate electrode since the 1980s due to its compatibility with silicon processing and its superior interface quality with the gate dielectric. Ideally, the work function difference between the polysilicon gate and the silicon substrate is zero, which gives it the primary advantage in usage. But, in the deep sub-micron regime, the advantages of polysilicon are diminished due to problems like poly-depletion, dopant diffusion (more acute when doping is higher), high leakage current, etc. Also, the series resistance of polysilicon is increasing with scaling. It is desirable to maintain low series resistance; otherwise the RC (resistance-capacitance) delay in a circuit becomes large and limits the speed of the device. Due to the low work function, polysilicon is suitable to be used as a buried channel device, but 'punch through effects' and 'drain induced barrier lowering' in the silicon channel ultimately limit the amount of achievable scaling. In addition, 'image charge' formed within the channel and appearing as a quantum capacitance (1-2 Å) has to be added to the effective inversion oxide thickness (T_{inv}). To solve the above mentioned problems alternative gate electrodes including metal gates were proposed[21,22,23,24].

The move to metal gates (MG) led to another challenge. The historical heavily-doped-polysilicon gates always provided a work function which was at band edge (4.1eV for NFET and 5.2eV for PFET). Metals, however, each have their own work function that is not necessarily at band-edge. To further complicate things, these metal work functions depend on the metal thickness and the process flow anneal cycles. The anneal impact, which typically drives a metal's work function from band-edge toward mid-gap, played a key role in decisions to move from a 'gate first' technology, where the metal gate is formed prior to the source/drain activation anneal, to a 'gate last' or 'replacement metal gate' (RMG) technology. In RMG a dummy polysilicon gate is built first, then, after the S/D anneals, the polysilicon gate gets replaced by the real metal gate[25]. At the same time the industry was moving to MG to eliminate the poly depletion problem, it also had to address the problem of tunneling leakage through the gate dielectric. In early 2000's, the conventional Silicon Oxynitrides (SiON) dielectric at ~1.2-1.7nm thickness [Intel at ~1.2nm[26] and IBM at ~1.7nm[27]] was leaking up to 30% of the channel current. The transition from SiO_2 (k~3.9) to SiON (k~4.5) in the 130nm node and then to HfO_2 (k~20) at the 45nm node provided similar or even improved capacitances at gate dielectric thicknesses large enough to essentially eliminate the leakage problem[28,26]. Since it was discovered that polysilicon undesirably reacted with the high-k (HK) dielectric the semiconductor industry was forced to change both the dielectric and the electrode at the same time. In general, continued scaling of the HK/MG systems is limited by reliability concerns[29]. Research work continues in hopes of finding a material capable of higher dielectric constants to increase C_{ox} (as in Equation 1) without affecting leakage or reliability.

Another revolutionary alternate approach to improving gate capacitance is to introduce a ferroelectric (FE) dielectric that can provide a negative capacitance (NC) effect[30]. NC FETs promise higher performance (i.e. larger drain current) at lower power (i.e. low V_{dd}) by decreasing the subthreshold slope (SS) below 60 mV/decade. Thus, NC transistors are categorized as steep sub threshold slope devices[31]. The NC effect has been proven on both planar[32,33] and non-planar[34] transistors using ferroelectric materials such as undoped[35] and doped hafnium oxide[36], hafnium zirconium oxide ($HfZrO_2$)[37] and lead zirconate titanate $[Pb(Zr_{0.2}Ti_{0.8})O_3]$[38]. Where hafnium based dielectrics must be in the orthorhombic phase to showcase ferroelectric characteristics[39], there are also reports on negative capacitance using poly dielectric on 2D materials[40]. It is also shown that ferroelectricity decreases as the thickness of the material is reduced[41]. For a logic device, a ferroelectric material with near zero hysteresis may be required which is challenging because the ferroelectricity of these materials is dependent on their thickness with larger thickness leading to higher ferroelectricity). The minimum thickness that has been proven to have ferroelectricity for HfO_2 is around 5nm while the gate dielectric thickness in current technology nodes is less than 2nm and is required to go to less than 1nm in next generation technology nodes. This means that even though the NC effect has been proven in the research phase at a very relaxed gate pitch, it is challenging to introduce the material into the next generation technology nodes because of the rigorous gate pitch scaling requirements.

Referring back to the Equation 1, another parameter that can increase the performance at lower power is the mobility of the channel[42]. Higher mobility means higher performance and lower power required to drive the transistor. The mobility of the channel carriers can be significantly changed by stressing the Si crystal, with tensile stresses enhancing electron mobility, and compressive stresses enhancing hole mobility[43]. Stresses can either be introduced by external films wrapping around the device (typically nitrides which, when they cool from deposition temperatures, exert their tensile or compressive stress)[44,45,46] or by internal materials changes by introducing high concentrations of elements which are larger-than-Si, such as Ge[43] or smaller-than-Si, such as C[47] to exert compressive or tensile strains. Both of these approaches suffer from the problem that, as pitches shrink, the amount of stressing material reduces which diminishes its ability to effect the desired change[48]. For example, a smaller 'gripping area' for an external nitride film limits its ability to stress the underlying drain region[49].

Mobility can also be altered by simply changing the orientation of the FETs with respect to the crystal plane. This is typically accomplished by keeping transistors in north-south orientation on the wafer, but by using a wafer with the crystal orientation rotated by 45 deg. The orientation of the surface normal (from channel to dielectric to electrode) also has an impact upon carrier mobility[50,51]. There is preferential surface orientation for both electrons and holes and it is not the same. Electrons and holes have higher mobility along the (100) and (110) crystal orientation respectively. The optimum mobility for both electrons and holes is along the (111) plane. For finFETs with the fin sidewall not in the same orientation as the wafer normal, this immediately comes into play. The percentage of stress effect is also dependent on the crystal orientation and thus mobility is affected by orientation. This means the sensitivity of mobility to stress in an embedded source drain is also dependent on the fin crystal orientation.

Beyond stressing and orientation, a final way to modify mobility is to change the channel material itself. Table 1 shows the fundamental bulk material properties of potential alternate channel materials[52,53,54,55]. The bulk properties provide attractive alternatives for device engineers.

Table 1. Bulk properties of key alternate channel materials.

Material/Properties	Si	Ge	InAs	InSb
Electron Mobility (Cm²/V.s)	1400	3900	40000	77000
Hole Mobility (Cm²/V.s)	450	1900	500	850
Bandgap (eV)	1.12	0.66	0.35	0.17
Lattice constant (Å)	5.431	5.658	6.058	6.749
Dielectric Constant	11.7	16.2	15.2	16.8
Melting point (°C)	1414	938.2	942	527

The two primary alternate channel materials under consideration are silicon-germanium ($Si_{1-x}Ge_x$) and indium gallium arsenide ($In_{1-x}Ga_xAs$) where x ranges from 0 to 1. Germanium (Ge) or alloyed mixtures of silicon and germanium (SiGe) lead to improved mobility up to ~2x, but also cause complexities in processing and increases in defect densities (i.e. dislocations induced by lattice size differences). Much larger improvement in mobility (10x or more) are possible by moving to group III-V materials. Binary III-V compounds such as indium arsenide (InAs) and indium antimonide (InSb), though theoretically promising to provide very high ON current, are less likely candidates for consideration due to very small band gaps and poor lattice compliance with silicon.

Though not insurmountable, the introduction of alternate channel materials brings considerable integration and manufacturing challenges[56]:

(1) High volume manufacturing (HVM) prefers to use larger wafers since they amortize process cost across more chips. The primary integration challenge of all alternate channel material is their absence of large wafers. This forces monolithic integration of alternate channel materials through epitaxial (epi) growth mechanism onto the silicon wafers as the only viable option. However, due to the very large lattice mismatch (4% for Ge, ~10% for InAs and ~19% for InSb) it is not easy to grow alternate channel on silicon.

(2) Difficulty in doping ions of opposite polarity. Unlike silicon, all alternate channel materials have preferred doping polarity. For example Germanium favors P type doping while Indium Gallium Arsenide (InGaAs) favors N type. This can also be evidenced from the very poor N to P mobility ratio in each of these channel materials.

(3) Poor gate dielectric interface that leads to higher interface defects. These interface defects act as scattering centers reducing the mobility of the channel drastically.

(4) Thermal mismatch between the material and silicon.

(5) Mobility in short channels substantially decrease from their long channel counterparts.

(6) Mobility also decreases as the width of the channel material decreases; for example, mobility substantially decreases from 15 to 5nm fin width[57].

(7) Dislocation dependent mobility for short channel devices is substantially lower than bulk mobility[57].

(8) The group III-V manufacturing ecosystem, including environmental safety and health (ESH) is still not matured enough to support HVM due to a lack of high through-put tools.

Despite these fundamental challenges there has been tremendous research done on these materials seeking to improve the above characteristics. As strained materials not only provide higher mobility but are also the least defective material when grown on lattice mismatched substrates, one particular integration approach has been to grow strained silicon on relaxed silicon germanium buffer layers[58] and strained silicon germanium on bulk silicon substrates[21,59]. Using these techniques it is fairly easier to bring alternate channel planar transistors to comparable performance as non-planar finFETs. There are also reports of introducing SiGe at the 28nm node where the work function of the channel is modulated by adjusting the Ge concentration[60]. It is easier to introduce Ge into extremely thin (fully depleted) silicon on insulator structures through condensation of SiGe to form higher concentration Ge channel[61,62]. Such channels have demonstrated the lowest number of defects and have shown improved performance.

2.2. *Source and drain contact engineering*

The rapid increase in parasitic series resistance is the primary source-drain and contact engineering challenge as pitch is aggressively scaled. Parasitic resistances in the source and drain regions act to limit the net voltage across the channel and therefore present a power concern. These parasitic resistances can be attributed to five components: (1) spreading-resistance under extension to gate overlap (2) spreading resistance under the spacer regions (both 1 and 2 are caused by lateral extension doping induced abruptness), (3) sheet resistance of extension and drain, (4) contact resistance of silicon-to-silicide, and (5) contact resistance of silicide-to-contact. Process innovations from the 1990s to 2000's were focused on improving the first three, thereby creating very sharp junctions and high active doping concentrations. These improvements were obtained by enhancements in annealing techniques obtained from rapid thermal anneals (5-30 seconds duration) to spike anneals (1 second) to flash or laser anneals (milliseconds). In essence the implant places the dopants where they are needed and then anneal simply moves them several lattice spaces to drop them into substitutional sites and become electrically active. However, in aggressively scaled channels, a lateral doping abruptness of less than 5nm/decade is required to maintain sharp S/D junctions and this poses a major processing challenge. In addition, because the implanted areas are becoming so small, the required number of dopant atoms in a given region (such as the halo on the drain side) becomes small enough that even small variations become significant enough to affect device behavior. Thus, random dopant fluctuation (RDF) is another major problem in aggressively scaled devices[63]. Beyond these challenges, with aggressive pitch scaling, innovations for reducing the 4th and 5th components, contact resistance, have become the focus for research.

Contact resistance (R_{cont}) is defined by the Equation 2.

$$R_{cont} \approx \frac{1}{L_{cont}} exp \left\{ \frac{4\pi\sqrt{\varepsilon m*}}{h} \frac{\phi_B}{\sqrt{N_D}} \right\} \tag{2}$$

where L_{cont} is the contact length, ε is the permittivity of the semiconductor, $m*$ is the effective mass of the semiconductor, h is the Planck's constant, ϕ_B is the Schottky barrier height between the diffusion region and metal contact layer, and N_D is the active dopant concentration in the semiconductor.

While the good thermally formed atomic bonds at the silicide-to-doped silicon interface eliminate most of the drawbacks of a plain metal to semiconductor interface, there is still room for improvement. The doping term, N_D, is easiest to modify by adjusting implant doses and energies. But more complex schemes are being utilized to reduce R_{cont}, involving the use of additional dopants of different sizes to compensate for lattice strain induced by the primary dopant. An example of this approach is co-doping with large-size In or Ga to overcome the tensile strain induced by high concentrations of small-sized B, thereby enabling net higher active doping concentrations[64]. Regarding the barrier height, all technologies to date have used a mid-gap silicide, meaning a roughly equal ~0.5eV jump for either electrons or holes to reach their respective band edges. A likely next step is to use dual silicides, one for NFET such as ErSi and a different one for PFET such as PtSi, each with a work function close to the respective conduction or valence band-edge, to minimize the barrier (~0.1eV) seen by an electron or hole as it jumps from silicon to silicide[65]. So far, the integration complexity introduced by running dual silicides has not been worth the benefits, but that may change in the near future[66].

One item in Equation 2 that is typically assumed to be unassailable is the contact length which is unavoidably shrinking in every technology generation. As transistors are packed tighter, very little room remains between the spacers of adjacent transistors to accommodate larger contacts. FinFETs have additional challenges in that they require tall contacts, leading to large source to drain capacitance, (C_{sd}). However, contact designs such as rounded S/D regions can reduce the contact resistance. A self-aligned contact (SAC) which involves a protective dielectric over the metal gate to prevent contact-to-gate shorts when the contact is misaligned and partially overlaps the gate is a necessity for aggressively scaled contacted poly pitch or gate pitch (CPP). SAC integration schemes are required to minimize any yield loss due to misalignment. Contact misalignment can lead to higher resistance and this is a primary integration and lithographic patterning challenge. As we move from planar to finFET technologies, S/D formation has become highly challenging. With contact resistance (R_{cont}) becoming such a significant performance issue, integration teams are considering structures beyond finFET, such as vertical transistors, which by their nature offer significantly increased contact areas.

The above discussion has focused on improving, or at least maintaining, drive current, while reducing, or at least holding, OFF current. But this is a DC characteristic and device engineers cannot ignore the fact that they're making AC devices with transistors switching ON and OFF at extremely high speeds. This brings forward an important challenge to the transistor development: the parasitic capacitance. Most capacitances on a MOSFET are coupled to the gate length and the primary capacitance knob is the gate to contact capacitance. This leads to either thinning the spacer; which in turn leads to either gate-to-contact leakage or unacceptably high overlap of doping under the gate corner; or lowering the k-value of the spacer which is the much preferred technique. Moving from conventional nitride spacers (k~7) to low-k spacers[67,68] (k~5) has brought noticeable improvements in AC performance. Material development continues to further reduce the k-value of the spacer dielectric.

So far we have discussed most of the challenges associated with the FEOL technology in a MOSFET. The next session will briefly introduce the BEOL scaling challenges.

3. Challenges and Approaches for CMOS Interconnect Scaling

Interconnect, also called BEOL or chip wiring, fulfills several essential functions on any integrated circuit such as signal routing, clock distribution, power/ground distribution network, long distance communication. It must satisfy the PPACR performance metrics. Some of the performance hurdles relate to wiring area, latency, bandwidth and signal integrity or noise, and reliability. However, over the previous two decades the movement of data has taken more and more of the operating energy budget; and in some cases today 75% of power dissipation is due to communication in and out of the chip[69,70]. Clearly this is an issue for mobile applications (where this affects battery life), to datacenters and server farms (where this affects operating power and cooling costs). The interconnect session in Section 6 will discuss some of the challenges in more depth and also addresses what is being done to continue the BEOL scaling. Research on co-optimizing technology with design and architecture, such as optoelectronics integration to accommodate bandwidth requirements in multicore devices, or neuromorphic computing that may not need very fast data transfer through individual wires, may be the future of interconnect technology.

4. Challenges and Innovations towards CMOS Advanced Patterning

As has been highlighted in the previous sessions, for most of the first 30 years of semiconductor circuit scaling, continuous improvements in transistor density, as dictated by Moore's law, were achieved through dimensional scaling, enabled to a large part by improvements in lithography resolution. Governed by Rayleigh's resolution criterion:

$$R = k_1 \lambda / NA \tag{3}$$

The aim of the semiconductor industry was to reduce the wavelength (λ) and increase the numerical aperture (NA) of the lithography tools at a pace sufficient to improve the resolution of critical feature pitches by 70% every two years while maintaining a k_1 factor (a process dependent constant) larger than 0.65. Maintaining a sufficiently large k_1 ensured high patterning limited yield (PLY) and allowed design rules to scale consistently from one technology node to the next. This made it possible to migrate existing circuit designs over many technology generations. The convenient life of simple dimensional scaling came to an end when wavelength improvements in lithography tools stopped at $\lambda = 193$nm while the semiconductor industry continued to scale at a relentless pace. The need to ensure reliable yield at ever more challenging resolution gave birth to the new engineering discipline of 'Computational Lithography'. Charged with developing increasingly complex resolution enhancement techniques (RET), computational lithographers allowed the semiconductor industry to maintain its aggressive pace of scaling in spite of inadequate improvements in fundamental exposure tool resolution. The price designers had to pay for pushing deeper into the 'sub-resolution domain' was a continuous erosion of their layout freedoms through ever tightening design rules. The scaling investment on the design side was matched on the process side by steadily increasing process complexity. Ensuring that both parties, the designers and the process technologists, contribute equally to the scaling effort became the responsibility of design-technology co-optimization (DTCO). The journey into the sub-resolution scaling domain is illustrated in Fig. 6[71].

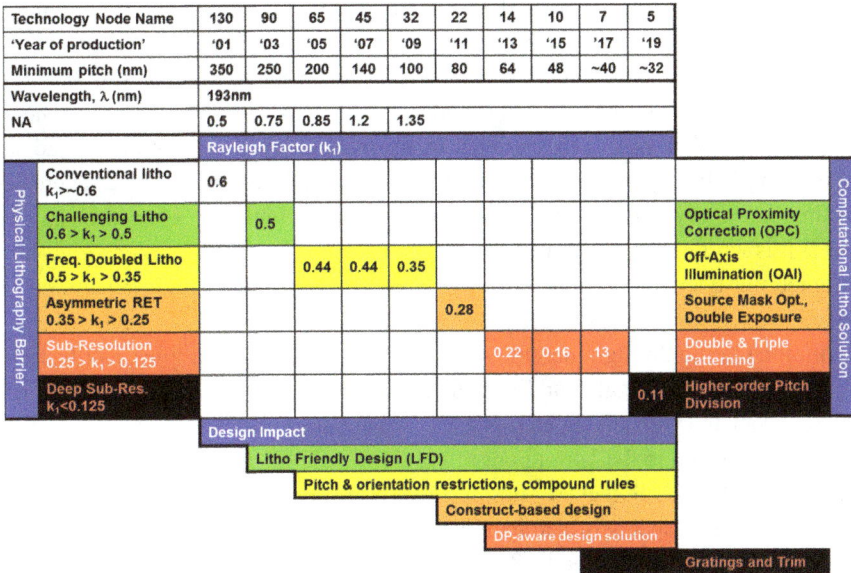

Technology Node Name	130	90	65	45	32	22	14	10	7	5	
'Year of production'	'01	'03	'05	'07	'09	'11	'13	'15	'17	'19	
Minimum pitch (nm)	350	250	200	140	100	80	64	48	~40	~32	
Wavelength, λ (nm)	193nm										
NA	0.5	0.75	0.85	1.2	1.35						
Rayleigh Factor (k₁)											
Conventional litho k₁>~0.6	0.6										
Challenging Litho 0.6 > k₁ > 0.5		0.5									Optical Proximity Correction (OPC)
Freq. Doubled Litho 0.5 > k₁ > 0.35			0.44	0.44	0.35						Off-Axis Illumination (OAI)
Asymmetric RET 0.35 > k₁ > 0.25						0.28					Source Mask Opt., Double Exposure
Sub-Resolution 0.25 > k₁ > 0.125							0.22	0.16	.13		Double & Triple Patterning
Deep Sub-Res. k₁<0.125										0.11	Higher-order Pitch Division

Physical Lithography Barrier — *Computational Litho Solution*

Design Impact
- Litho Friendly Design (LFD)
- Pitch & orientation restrictions, compound rules
- Construct-based design
- DP-aware design solution
- Gratings and Trim

Fig. 6. Relating physical lithography barriers to computational lithography solutions and associated design implications through the Rayleigh factor (k₁) [Source: Reprinted with permission from Reference 71, Copyright 2016 SPIE, doi:10.1117/3.2217861].

In the k₁ regime between 0.65 and 0.5 at around the 90nm technology node, the substantial loss of image fidelity had to be compensated for by optical proximity correction (OPC). The advent of elaborate though imperfect post design layout manipulations used in OPC, introduced the design community to the concept of lithography-friendly design (LFD). LFD aimed at identifying rogue layout configurations that did not violate any particular set of design rules, but for a variety of reasons posed a yield risk (inability to print / etch). Initial adoption of what became known as 'hotspot detection and repair' was slow, but perhaps more importantly the concept of LFD for the first time opened up channels of communication between the previously isolated design and process communities. This close collaboration was instrumental in the technology nodes that operated at a k_1 below 0.5 requiring strong, 'frequency doubling' resolution enhancement techniques (RET) and forcing a sharp rise in the number and complexity of design rules. This k_1 domain, spanning the 65nm to 32nm technology nodes, was extended by the introduction of higher NAs in immersion lithography and allowed designers ample time to gain proficiency at complex design rules such as 'width depended spacing rules', 'short edge rules', and 'forbidden pitch rules' before tackling the next k_1 hurdle toward the 22nm node. Often underappreciated in its significance, the drop below k_1 of 0.35 forced lithographers to use asymmetric off-axis illumination and introduced the design community to 'preferred orientation design rules' and even ventured into 'double patterning rules' made necessary by the use of line-end cut patterning approaches. Collectively, the design restrictions necessary to keep moving past the ultimate lithographic resolution limit of $k_1 = 0.25$ would have paralyzed a designer coming back into the industry after a five node sabbatical. The only reason 'single orientation', 'fixed pitch', and 'construct-based' prescriptive design rules were survivable was the gradual elimination of designer's freedom and the increased collaboration between designers and process engineers that was fostered over many technology nodes. While in advanced technology nodes, lithography often has to take the brunt of the blame for the severe design restrictions necessary to support extremely complex multiple exposure patterning processes, it is worth noting that, had the device engineers tried to introduce high-k metal gate strained finFET devices in the 90nm technology node, the design implications would have been seen as completely unsupportable. Only through the many years of enabling sub-resolution scaling through incrementally more restrictive design rules, as illustrated in Fig. 7, did unidirectional gates at fixed pitch, limited gate lengths, and discrete device widths, all associated with advanced transistor architectures, stand a chance of adoption for volume manufacturing in the 14nm node.

The degree to which severe design restrictions have become an indispensable component of many aspects of semiconductor design as well as manufacturing is made evident in the 5nm technology node. The much delayed introduction of extreme ultraviolet lithography (EUVL) finally provides a long awaited resolution boost but is met with no desire to relax design restrictions originally implemented for the sole benefit of lithography, as will be shown in the next section (4.1) of this paper.

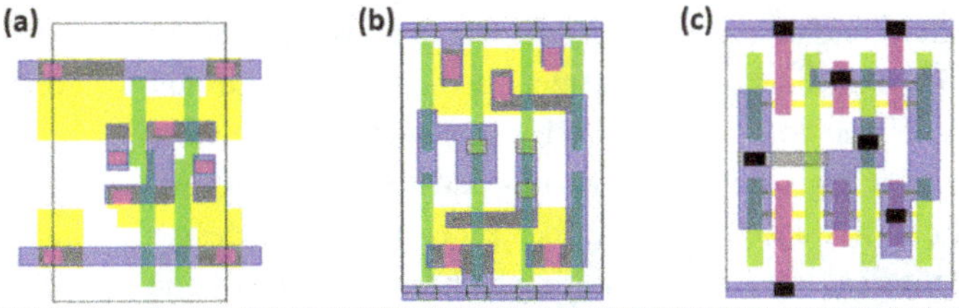

	130nm Node	22nm Node	14nm Node
Diffusion	Complex polygon with internal routing	Simple rectangle, source/drain only	Discrete unidirectional fins
Contact	1 level, all squares	2 levels: diffusion vs poly contact, bars and squares	3 levels: diffusion vs poly local interconnect plus square vias
Poly	Bidirectional, variable pitch	Unidirectional single pitch, dummy poly	=
Metal	Complex, bidirectional	=	=

Fig. 7. The gradual tightening of the design space as a consequence of lithography friendly design enabled the transition to advanced devices like finFET. Figures 7(a), 7(b) and 7(c) corresponds to 130, 22 and 14nm technology nodes.

4.1. *Complementing pitch reduction with DTCO-based scaling*

To help the more process or technology oriented readers appreciate the impact of DTCO, a simplified view of a standard cell logic design flow is shown in Fig. 8.

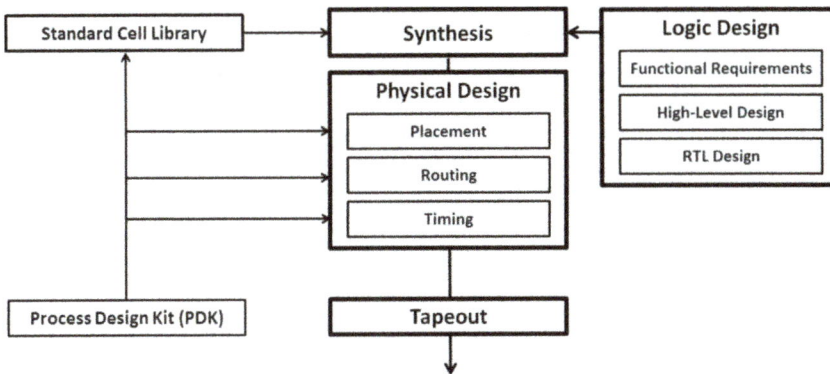

Fig. 8. Main elements of a highly simplified the standard cell design flow identifying three major blocks: logic design, standard cell library, and physical design.

The basic operation of this commonly used standard cell design flow can be described as three major blocks:

The 'logic design' block where the functional design information is created communicated to the synthesis tool in algorithmic form using register transfer language (RTL). The synthesis tool maps the algorithmic description of the design into a collection of standardized Boolean logic functions such as NAND gates (not and), NOR gates (not or), or AOI gates (and or invert) and storage functions such as latches. To allow logic designers to focus on optimizing the functional design without having to worry about the physical rendering of the logic functions for any particular technology node or semiconductor foundry, these logic functions are pre-rendered in layout form (i.e. polygon drawings) in a standard cell library. Separating the 'functional logic design', which is product specific, and the 'physical layout rendering', which is technology node and foundry specific, allows functional designs from a fabless design house to be ported from one technology node to the next and from one foundry to the other. It also means that functional logic designers have, for the most part, been excluded from DTCO efforts to date. Once mapped into a layout description by selecting the appropriate cells from the standard cell library, the design enters the physical design block where the placer arranges the cells into logic blocks based on their electrical vicinity and the router wires the cells to establish the power distribution logic signal flow. The final step in the physical design flow before the chip is taped-out or released to the foundry is timing closure where the design is fine tuned to ensure all signals traversing parallel logic paths arrive at the designated latch in the time allotted by the clock frequency of the logic block.

While the functional logic design is deliberately shielded from the details of the manufacturing process, the capabilities and constraints of a particular technology node are communicated from the semiconductor foundry to the various stages of the physical design flow via the process design kit (PDK). The PDK contains design rules and electrical models necessary for the IP providers (i.e. in house or 3rd party standard cell library creators) to create and verify the standard cell library and for the electronic design automation (EDA) engineers to build the appropriate place and route (PnR) solution. This brief introduction to digital standard cell design highlights the complexity of DTCO. On both sides of the collaboration it is impossible to have one individual represent all the concerns and optimization objectives since so many complex technology elements require many specialized subject matter experts. The rest of the session will discuss DTCO approaches used to complement critical feature pitch reduction to achieve adequate area scaling in nodes beyond 14nm.

4.1.1. *The scaling roadmap*

To put the discussion of logic scaling beyond the 14nm technology node (N14) into a meaningful context the hypothetical but realistic technology progression roadmap shown in Table 2 is used throughout this discussion. In Table 2 each technology node is characterized by three critical feature pitches:
(1) The wire pitch describes the smallest width and space combination allowed for the lowest metallization levels and is primarily responsible for cell height scaling.
(2) The poly pitch, also known as the gate pitch, describes the tightest transistor placement and is primarily responsible for the cell width scaling.
(3) The fin pitch describes the pitch of active channels in a finFET device and is a strong contributor to performance scaling.

In this scaling roadmap, the poly pitch is set at a 5-to-4 pitch ratio relative to the metal pitch and the fin pitch is set at a pitch ratio of 3-to-4. These are simply illustrative values and there is no fundamental reason for these exact pitch ratios, however, design efficiency is improved in advanced technology nodes dominated by gridded layout styles by maintaining clean pitch ratios.

Table 2. A hypothetical but realistic roadmap for N14 to N5 scaling (the abbreviations N14, N10, N7, N5 corresponds to 14, 10, 7 and 5nm technology nodes).

Hypothetical Roadmap					
Node Name	N14	N10	N7	N5	
Wire Pitch (nm)	64	48	40	32	
Poly Pitch (nm)	80	60	50	40	5:4
Fin Pitch (nm)	48	36	30	24	3:4

In addition to providing some dimensional context to the often arbitrary node names, Table 2 highlights some of the more prominent scaling inflection points:

(1) 48nm wire pitch is approximately the limit at which 193nm immersion lithography can resolve bidirectional patterns. This is achieved by interdigitating two or more patterns at $\geq$80nm pitch in a sequence of lithography and etch operations. Appropriately names litho-etch-litho-etch (LELE) this multiple exposure patterning technique is limited by the alignment error between the interdigitated features.

(2) 40nm wire is the well accepted limit of single orientation patterning achieved by enhancing 193nm immersion lithography with self-aligned double patterning (SADP), a sidewall deposition based frequency doubling RET.

(3) 40nm poly pitch is seen as the electrostatic and manufacturability limit of finFET devices. At the gate length necessary to reliably turn off the tri-gate finFET, 40nm pitch becomes the limit at which all the sidewall spacers and source/drain contacts can be reliably deposited with acceptable variability.

(4) 24nm fin pitch is the approximate manufacturability limit for fins in a finFET device. Mechanical stability of the high aspect ratio fins, the ability to deposit sufficient work-function metal and low resistance metal into the space between fins, as well as the ability to cut unwanted 'dummy fins' out of the patterned fin array all become limiting factors preventing further fin pitch scaling.

Figure 9 shows the challenging resolution domain in which these technology nodes take place. For the wiring pitches listed in Table 2, the lithography complexity increase, as expressed by a steady reduction in k_1 factor, is shown in Fig. 9. The technology nodes being discussed in this section reside well below the single expose limit of state-of-the-art lithography tools and could even penetrate below the double exposure limit in N5.

Lithography Complexity (k₁) by Node

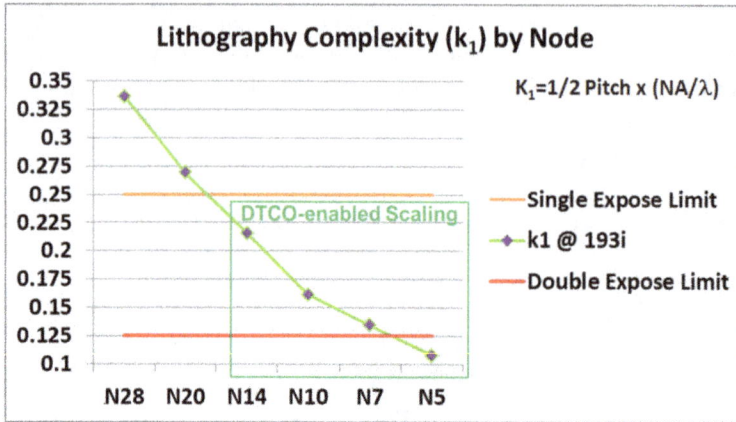

$K_1 = 1/2$ Pitch x (NA/λ)

DTCO-enabled Scaling

— Single Expose Limit

◆ k1 @ 193i

— Double Expose Limit

N28   N20   N14   N10   N7   N5

Fig. 9. Lithography complexities for the nodes of interest in this paper. The k_1 factor for each node's wiring pitch shows that the N14-N5 nodes, the primary domain of DTCO-enabled scaling, take place at very challenging resolution.

The central message of this section is illustrated in Fig. 10. Due to scaling inflection points, as the ones outlined above, attempting to achieve the desired node-to-node area scaling through pitch scaling alone would result in significant cost increase and schedule risk. It is therefore more desirable in advanced technology nodes to complement pitch scaling with DTCO enabled cell size reduction, as will be shown in sections 4.4-4.7.

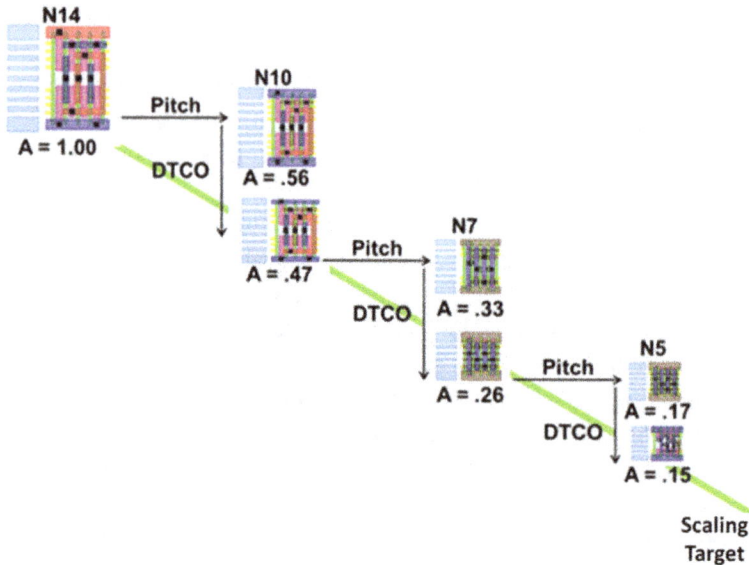

N14
A = 1.00

Pitch

DTCO

N10
A = .56

A = .47

Pitch

DTCO

N7
A = .33

A = .26

Pitch

DTCO

N5
A = .17

A = .15

Scaling Target

Fig. 10. To achieve the desired 50% node to node area scaling advanced technology nodes rely on a combination of pitch scaling and DTCO facilitated cell area reduction. The abbreviations N14, N10, N7 and N5 correspond to the 14, 10, 7 and 5 technology nodes respectively.

4.1.2. *AOI–DTCO's Canary cell*

In addition to looking at actual numerically anchored dimensional inflection points, it is instructive to look at a real standard cell logic layout when discussing scaling challenges. The logic cell used in Fig. 10 and in the following discussion to illustrate a number of key DTCO principles is shown in Fig. 11 as it might be drawn in the N14 technology node. It represents the 'and-or-invert' (AOI) logic function which was introduced in section 4.1. The critical levels used to render this cell layout are listed to the left of the cell while its logic truth table is shown to its right. One sample logic path is highlighted in the table and traced in the layout to illustrate how the three transistors and associated signal wiring are used to form the desired output. From an overall functional design standpoint, there is nothing unique about this AOI, but it is a nice logic cell to use for DTCO discussions since it is simple enough to be clearly rendered in a small figure yet complex enough to stress the patterning and manufacturing capabilities. As was already shown in Fig. 7, several nodes of sub-resolution patterning have forced the diffusion (here formed by fins), local interconnects, and poly levels to be rendered in a highly restricted gridded layout style. The only design level shown in Fig. 11 that maintains a significant degree of layout freedom is the 1st metal level. This highlights an important aspect of DTCO that will be further discussed in section 4.1.3. In some cases the optimal tradeoff between layout simplification and process complexity works out in favor of maintaining more complex layout geometries. The 1st metal level in this popular cell architecture serves several critical functions: it forms power-rails that run continuously along the horizontal cell boundaries to form a robust low resistance power delivery network (PDN), it also forms input pins (labeled 'A-C' in the table of Fig. 11) that run perpendicular to the power rails and provide the router with access points to wire the cell, the output pin (labeled 'Out' in the table of Fig. 11) is formed from bidirectional 1st metal that allows it to collect PMOS and NMOS

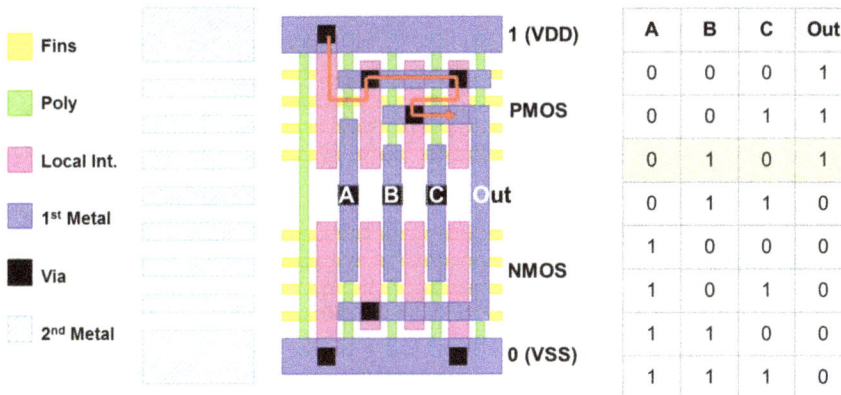

A	B	C	Out
0	0	0	1
0	0	1	1
0	1	0	1
0	1	1	0
1	0	0	0
1	0	1	0
1	1	0	0
1	1	1	0

Fig. 11. AOI cell: poly (green), fin (yellow), tungsten strap (pink), contacts (black), 1st metal (red and blue), and 2nd metal tracks for routing (light blue, left of the cell). The line highlighted in the logic truth table is traced in the cell to show the signal flow from the power-rail to the output pin.

signals and wire them to a long vertical pin that the router can connect to, finally 1st metal is also used to form any source-drain connections that are required to render more complex logic functions. Restricting 1st metal to be unidirectional, as will be seen in section 4.1.3, comes at the cost of having to provide additional wiring resources to complete all these necessary functions.

Keeping in mind from Fig. 9 that the N14 technology node is already deep in the sub-resolution domain and rely on multiple exposure patterning to achieve its dimensional targets as well as finFET to achieve its device performance targets, the following sections focus on scaling beyond N14.

4.1.3. *N14 to N10 scaling*

The scaling from N14 to N10 illustrates how, unlike design for manufacturability (DFM) where designs are optimized to improve manufacturability, DTCO often results in process complexity increase in favor of maintaining design ability. As shown in Fig. 12, to scale the established cell architecture to the N10 density target, a third exposure has to be added to the 1st metal (M1) patterning. This is not driven by raw pitch resolution as the 48nm wiring pitch of N10 can be resolved in two exposures, but rather by the need for small tip-to-side spaces that are essential for the bi-directional M1 which is needed in this highly efficient cell architecture. Since the lithographic interaction distance (indicated in grey diagonal hatch in Fig. 12) between features remains constant as feature pitch decreases, more features interact (as indicated by the red markers on the right of Fig. 12) and need to be patterned by different masks (as indicated by the green, blue, and purple coloring in Fig. 12). To avoid color conflicts without relaxing tip-to-side space, a 3rd color had to be introduced for 1st metal in N10.

N14   **N10**

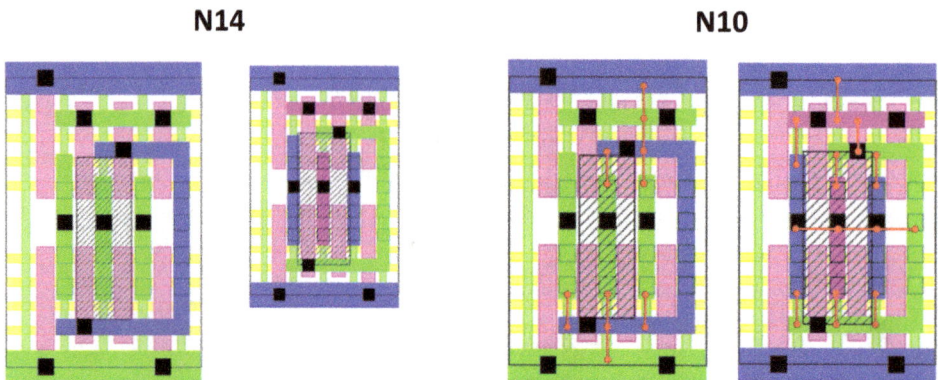

Fig. 12. AOI cells in N14 (left) and N10 (right) shown to scale (left) and at equal cell height (right). The colors in Fig. 12 represent the same colors as in Fig. 11.

Adding a 3rd exposure to the 1st level metal of the N10 node introduces not only additional process cost but also adds design complexity and cost. Designers and design rule checking tools have to understand how to resolve three color mapping conflicts while placement tools have to learn how to take advantage of the additional degree of freedom though color-aware placement solutions. To compensate for this additional process and design cost, and taking advantage of the smaller tip-to-side space afforded by the 3rd metal color as well as the higher than necessary drive current on finFET, careful design rule and layout optimization allows cell height scaling from 9T (with 4 NMOS and 4 PMOS fins) to 7.5T (with 3 and 3 fins). This DTCO effort brings the total area scaling to 0.47x (0.56x from pitch scaling and 0.83x from cell height scaling).

4.1.4. *N10 to N7 scaling*

The N7 node pushes optical lithography resolution so close to the fundamental physical limit that extensive DTCO is needed to enable grating-and-cut patterning processes. The bidirectional M1 pattern shown in Fig. 13(a) cannot be resolved by grating-based techniques needed at these dimensions. Fig. 13(b)-(d) show the evolution of unidirectional metal cell architectures in which cell wiring is split into horizontal M0 and vertical M1, showing incremental optimization of pin access and transistor wiring as the design and process details of the local interconnect and wiring stack are co-optimized. The ample pin access afforded by splitting the bidirectional M1 into two horizontal M0 and vertical M1, enables careful design rule and layout optimization to further scale cell height from 7.5T in N10 to 6T in N7, yielding a combined pitch and cell height driven area scaling of 0.56x.

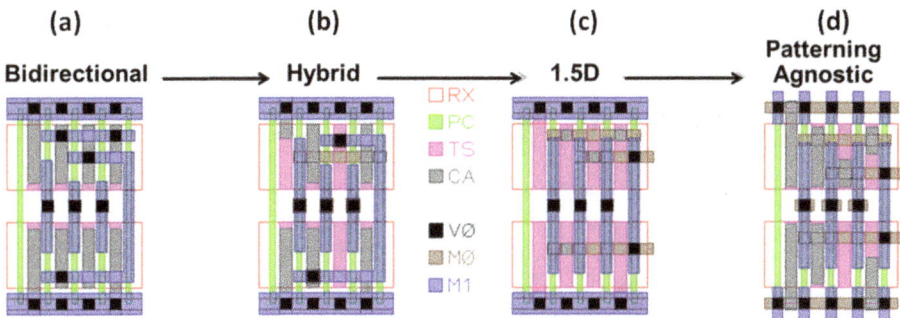

Fig. 13. (a) AOI rendered with bidirectional M1 (horizontal power-rail, vertical pins, horizontal signal wire), (b) hybrid cell layout introduces a second metal level (M0) to relieve tip-to-side spacing constraints but maintains bi-directionality, (c) 1.5d cell image moves all horizontal wiring except the power rail onto M0, opening up more patterning choices for M1, (d) fully unidirectional M1 allows grating based patterning techniques to be employed.

An important detail not addressed in this discussion is the power-rail implication of unidirectional metal. The traditional fully redundant stack of parallel metal wires is no longer feasible in grating-based patterning and alternative approaches that balance manufacturability and power-performance implications by exploring innovative cell architectures as well as new place-and route capabilities have to be explored in close collaboration with EDA tool providers.

4.1.5. *N7 to N5 scaling*

The N7 node is emerging to be a bitter-sweet node. As shown in Fig. 14, the long awaited arrival of EUVL theoretically provides substantial relief in resolution. This improved patterning resolution should open up the possibility of either relaxing the design restrictions at the target pitches for N5, or maintaining the design restrictions but over-scaling the critical pitches to achieve more area scaling. However, the integration solutions for the device and interconnect scaling challenges that will be explained in the following sections rely on very regular layout configurations and do not leave any room for more aggressive pitch scaling. While EUVL will provide substantial productivity improvements by collapsing complex sequences of multiple 193i exposures into a single EUVL exposure, layouts are likely to remain highly structured and pitch scaling will remain challenging.

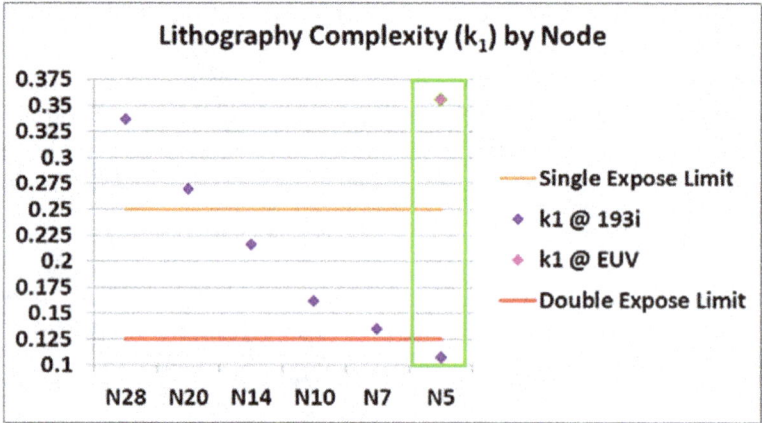

Fig. 14. Compares optical vs EUV lithography and demonstrates substantial productivity improvement with EUVL.

The next section (section 5 and 6) will discuss challenges associated in fabricating non-planar FEOL and BEOL transistor elements. In particular, section 5 will explain in detail the challenges associated in fabricating (5.1) the fin, (5.2) epitaxial source and drain, (5.3) dummy gate and spacer, (5.4) self-aligned contact and replacement metal gate, (5.5) diffusion break: single and double diffusion break, (5.6) source and drain contacts, (5.7) transistor architecture beyond finFETs. Section 6 will discuss the BEOL challenges such as (6.1) interconnect scaling, (6.2) reducing interconnect resistance, (6.3) reducing interconnect capacitance, (6.4) design and patterning and (6.5) packaging. Both these sections should enable the reader to gain an understanding of the complexities of integration and building production-worthy complex novel transistor structures.

5. FEOL Integration Challenges in Developing Non-Planar FinFET and Subsequent Technologies

Since the introduction of finFET technology at the 22nm node[72], continuous reductions in the dimensions of the transistors have managed to maintain the aggressive pace of scaling as dictated by Moore's law[73,74,75]. Table 3 shows the target dimensions from the 22nm to the 7nm node, and Fig. 15 shows a typical finFET fabrication process flow. A lot of new challenges in patterning and integration have been encountered with introduction of finFET technology and those challenges will become increasingly significant as the technologies further scale beyond the 10nm node.

Table 3. The technology target dimension from corresponding references. These numbers from the literature will be used for discussions in this section.

Tech. Node	22nm[72]	14nm[73]	10nm[75]	7nm[76]
Poly or Gate pitch (CPP)	90nm	70nm	64nm	54nm
Fin pitch	60nm	48nm	42nm	27nm

- Fin and Isolation formation
- Dummy Gate Deposition and RIE
- Spacer and EPI formation
- ILD and CMP
- Dummy Gate open and Removal
- HKMG Formation and Recess
- Self aligned contact (SAC) Formation
- MOL BEOL Metallization

Fig. 15. FinFET fabrication process flow.

A summary of the primary device performance challenges in finFET fabrication is described here. The fin formation process includes (1) fin patterning, (2) fin cut, and (3) isolation formation. The fin shape (i.e. width, height, and profile) and fin pitch (FP) are very important in driving the performance of finFET. The width of the fin (W_{eff}) is an important knob in enhancing the electrostatics of the finFET. Both 'drain induced barrier lowering (DIBL)' and 'steep sub-threshold slope (SS slope)' decrease as the width of the fin decreases. Therefor it is very important to reduce the width of the fin, the challenges of which are explained in the subsequent sessions. Also, an optimum aspect ratio between the 'gate length' and 'fin width' must be maintained to obtain the right gate electrostatics. The fin sidewall angle also impacts the finFET performance because mobility of the transistor depends on the crystallographic planes. As the sidewall angle increases the mobility decreases. Sidewall angle always degrades the drive current. However, there is a process advantage to having an angled fin. For short channel finFETs with uniform gate dielectric, the electric field in the gate dielectric is the smallest at the fin corner as compared to the top or sides of the fin. This means, the maximum current density is not in the top narrowest portion of the fin but is distributed towards the fin volume due to the quantum confinement. In other words, the rounded corners have the best electrostatics and the least amount of barrier lowering. Therefore, angled fins with narrow fin at the top have a reduced thermal stress leading to a reliability advantage. The parasitic capacitance drops significantly as the fin pitch is reduced, but fin pitch reduction presents a major patterning challenge. Parasitic capacitance also reduces with increased fin height; though increasing the fin height introduces major structural challenges and further complicates patterning. Also, there is a contact resistance penalty as we increase the fin height thus reducing drive current. In summary: considering all these trade-offs, reducing fin pitch is the most effective means of improving device performance.

In addition to fin geometry optimization, source and drain contact engineering plays a major role in device optimization. In aggressively scaled CPP the space between active gates gets so small that contact-to-gate shorts have to be prevented through the use of SAC. Contact geometry can also reduce the contact resistance. For example, rounded contacts may have lower resistance compared to square contacts. This is because contact misalignment can lead to higher resistance and can thus increase integration and lithography challenges. Contact misalignment is less affected by rounded contacts[77]. Thicker gate spacer can also give higher performance because of lower overlap capacitance but due to the limited space available in aggressively scaled CPP, it is difficult to fabricate a thick spacer. A low-k spacer can significantly reduce overlap capacitance and also reduce the power density requirement. However, low-k gate spacer processes present integration challenges associated with a new material being introduced to the technology, and because these materials tend to be less robust to typical processing (cleans, for example). Replacement Metal Gate (RMG) processes at small gate dimensions introduce challenges caused by the tight spaces into which materials must be filled. In addition, these processes have to be optimized to generate lower gate resistance. This session will now review the challenges and potential solutions based on the process flow in Fig. 15.

5.1. *Challenges in Fin formation*

5.1.1. *Fin patterning: From SADP to SAQP*

To achieve adequate active region at a given footprint and to meet the drive current requirements, a fin pitch of 60nm was selected for the 22nm technology node[72], and gradually scaled further for the 14nm and 10nm nodes. In the 7nm node the fin pitch will reach sub 30nm. These dimensions far exceed the resolution capability of 193nm wavelength immersion (193i) lithography. Extreme ultraviolet lithography (EUV) at a wavelength of 13.5nm may be an option to meet the resolution requirements.

Advanced resolution enhancement techniques, such as multiple exposure patterning where several sequential cycles of lithography and etch are needed to render a mask level or self-aligned patterning through sidewall image transfer can be used to overcome the resolution limitations of the 193i optical lithography. For reduced pitch walking (undesired alternating big-pitch/small-pitch) and improved fin CD control, self-aligned multiple patterning has been adopted for fin patterning.

For a fin pitch greater than 40nm, self-aligned double patterning (SADP) has been used. Fig. 16 shows the process flow for SADP fin process[78]. Fin CD is set by spacer CD, and each mandrel defines two fins, so the pitch of the fin is half that of the mandrel pitch. An additional mask can be used to define the diffusion regions for planar devices that are manufactured on the wafer together with the SADP patterned finFET. Pitch walking is controlled by minimizing the difference between space "a" and space "b". Space "a" is determined by the initial mandrel CD and the spacer CD losses during fin etch. Space "b" is equal to the fin pitch minus the spacer CD and space "a". Therefore, the most important parameters to set control, are the mandrel CD, spacer deposition thickness, and etch bias during the spacer reactive-ion etching (RIE) and final fin etch.

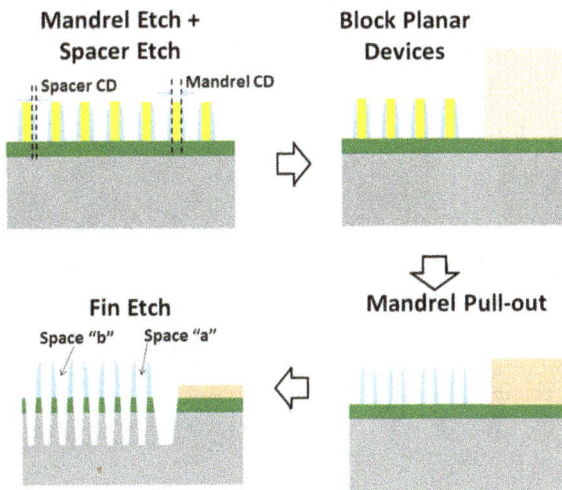

Fig. 16. Schematic drawing of a SADP fin formation process. Fin pitch is defined by ½ of initial mandrel pitch [Source: Reprinted with permission from C. Park *et al.*, Reference 78].

As fin pitch continues to scale to sub-30nm dimensions for the 7nm node and beyond, self-aligned double patterning can no longer achieve the required resolution and self-aligned quadruple patterning (SAQP) needs to be used. Alternatively, the same resolution could be achieved by interdigitating two SADP exposures but this would require 2 lithography processes which would add an overlay shift (~3nm) to pitch walking. Large pitch walking causes active fin height variation which impacts device drive current as well as source and drain epi formation. Thus, SAQP is a better option to maintain a tight pitch walking at the 10nm node and beyond.

(a)

(b)

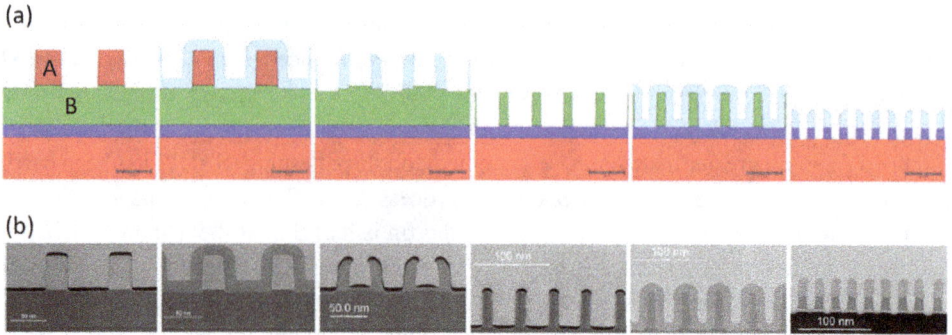

Fig. 17. (a) Simulation images of the stages of self-aligned quadruple patterning (SAQP), from left to right: patterning of the first core (brown) onto a mandrel (green); deposition of silicon dioxide (SiO_2) (light blue) by atomic layer deposition (ALD); etching of the first spacers; etching of the mandrel to produce the second core; further deposition of SiO_2 by ALD; and etching of the second spacers and silicon nitride pad (dark blue). The scale bars represent 30nm. (b) Transmission electron microscopy (TEM) images of the stages of SAQP show, from left to right: patterning of the first core onto a mandrel; deposition of SiO_2 by ALD; etching of the first spacers; etching of the mandrel to produce the second core; further deposition of SiO_2 by ALD; and etching of the second spacers and silicon nitride pad [Source: Reprinted with permission from E. A. Sanchez *et al.*, Reference 79/IMEC, Copyright 2016 SPIE, doi: 10.1117/2.1201604.006378].

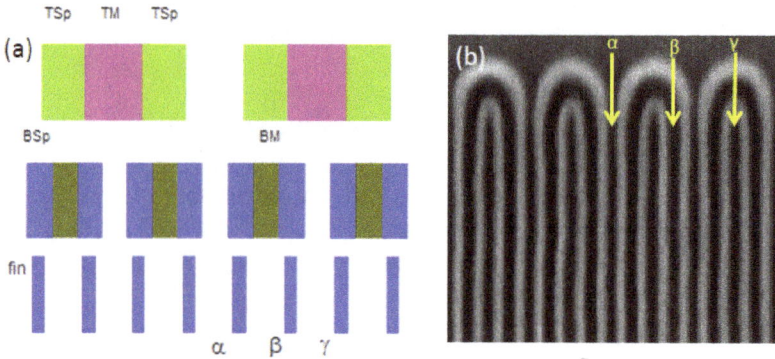

Fig. 18. (a) Schematic flow for self-aligned quadruple fin patterning (SAQP). α, β and γ corresponds to various pitches and spaces that need to be optimized for SAQP process; (b) Top-down scanning electron microscopy (SEM) images of the fins formed with SAQP process.

Fig. 17 shows a schematic and TEM flows of SAQP fin patterning[79]. The key idea is to have two layers of mandrel materials with different etch selectivity. By forming the 1st spacer on the top mandrel material and transferring the pattern into bottom mandrel material, the pitch is reduced to ½. Then a 2nd spacer will be formed on the bottom mandrel which further decreases the pitch by half.

As shown in Fig. 18, Controlling α, β, and γ spaces is very critical to minimizing pitch-walking. This requires co-optimization and tight control of multiple parameters and is much more difficult than SADP. SAQP has three process parameters that determine the pitch walking: top mandrel CD, top spacer thickness, and bottom spacer thickness. Since

SAQP has a larger number of variables to control, it is naturally a more difficult process to yield than SADP.

Figure 19 shows calculated fin pitch walking based on three SAQP process parameters. The y axis in the figure shows pitch walking as function of three process parameters (x axis represents top mandrel CD, each color represents different top and bottom spacer thickness). Fig. 19 shows that zero pitch walking is obtained for different values of the three parameters. It also shows the rapid increase in pitch walking by 2nm with a 1nm drift in the various process parameters.

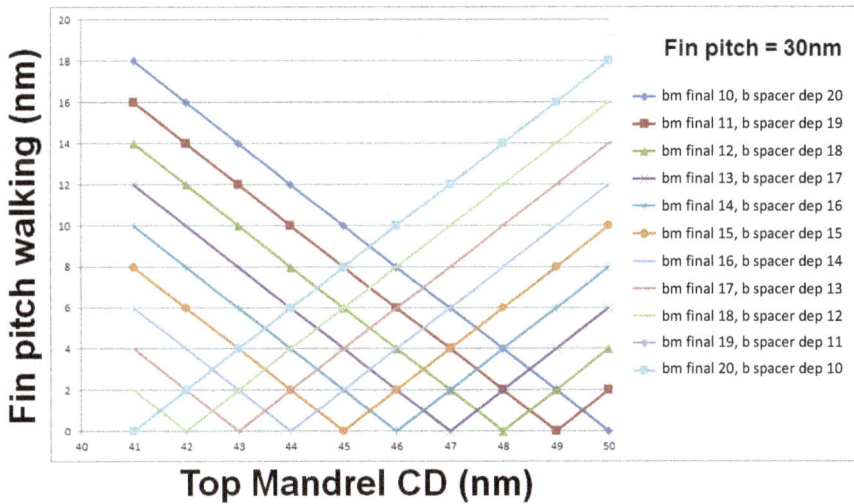

Fig. 19. Calculated pitch walking as a function of three process parameters. "bm final" and "b spacer dep" in the legend correspond to "bottom mandrel RIE CD" and "bottom spacer thickness" in nm, respectively.

Based on the data shown in Fig. 19, we can establish two rules to obtain zero pitch walking:
(1) Final bottom mandrel CD + final bottom mandrel space CD = 2x fin pitch.
(2) Bottom spacer oxide deposition thickness + Final bottom mandrel CD = Fin pitch.

Since SAQP scales the pitch four times, the first and second pitch walking steps will have to be tuned separately. Thus, there are four parameters to control in the SAQP process: three pitch walking parameters and the final fin CD. In addition to pitch walking, the fin profile also has to be controlled by achieving vertical top and bottom mandrel profiles and controlling the RIE bias during top mandrel RIE, bottom mandrel RIE and channel RIE.

Since the SAQP process is an extension of the SADP process, both these processes are identical up to bottom mandrel RIE. The top mandrel CD and top spacer thickness have to be controlled at the SADP level while the bottom spacer thickness that defines the final pitch walking is controlled at the SAQP level. It is difficult to recover from pitch walking defined by the bottom mandrel RIE, as discussed in Fig. 18. Therefore, the first pitch walking has to be tuned through iterations at the SADP before the second pitch walking tuning can be done. Fig. 20 further shows a flow chart of how to control the pitch walking.

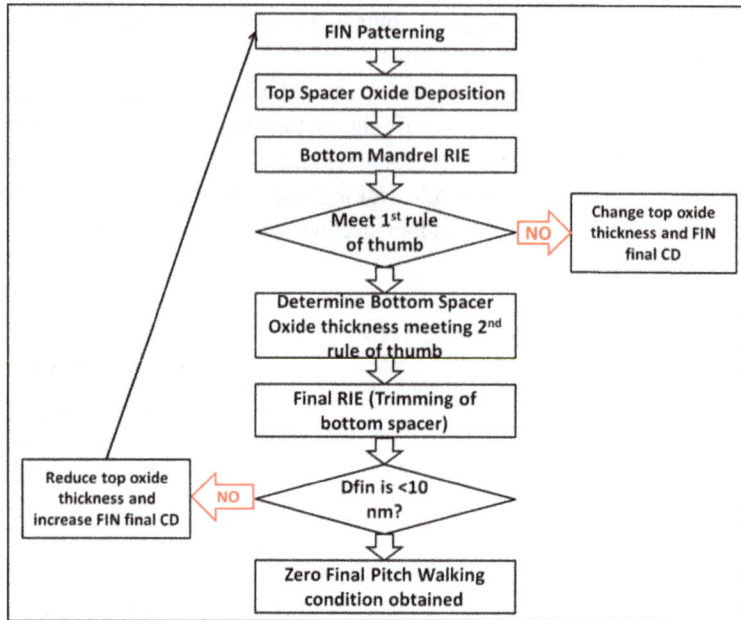

Fig. 20. Flow chart to handle pitch walking control. The figure shows the two rules of thumb that can be used to obtain zero pitch walking. (1) Final Bottom Mandrel CD + Final Bottom Mandrel Spacer CD = 2 X Fin Pitch (2) Bottom Spacer Oxide Deposition thickness + Final Bottom Mandrel CD = Fin Pitch.

5.1.2. *Fin cut*

Another challenge associated with finFET scaling is 'fin cut'. As the fin pitch scales to sub-30nm and SAQP has to be employed, it is best to pattern a continuous array of fins with minimal pitch variation. To achieve the larger fin space needed between different devices, 'dummy fins' have to be cut out of this continuous array of fins. At tighter fin pitches, the cut margin become seriously degraded, and it's very hard to completely cut the unwanted fins without any damage to the wanted fins considering all process variations.

Figure 21 shows schematics of 'Fin cut first' scheme vs 'Fin cut last' scheme[80]. The primary difference between these two methodologies is the point in the process sequence when the dummy fin is removed. The 'Fin cut first' scheme (see Fig. 21(a)) defines the cut on the fin hardmask level such that there is only one RIE step to form the final fin shape. This sequence makes the isolation process simpler by requiring only a single oxide void free 'gap fill' and planarization. On the other hand, the 'Fin cut last' scheme (see Fig. 21(b)), as the name implies, removes the dummy fins after the final fin image is etched into the substrate. This means, the fin cut has to be done after the isolation process, which requires an additional isolation process after the fin cut. Therefore, the fin cut last scheme requires a larger number of process steps than the 'fin cut first' scheme. As the fin cut is done across the full topography of the final fin structure in the 'fin cut last scheme', cut mask misalignment can lead to spikes of the residual dummy fins. The size of Fin spike is dependent on the RIE profile and the cut lithography overlay margin. The spike can be a problem if it sticks out from the isolation area, where unwanted epi can grow during source and drain formation. As the cut profile angle can't be a perfect 90 degree, the remaining fin spike becomes taller for thicker fin hardmasks and taller fin heights.

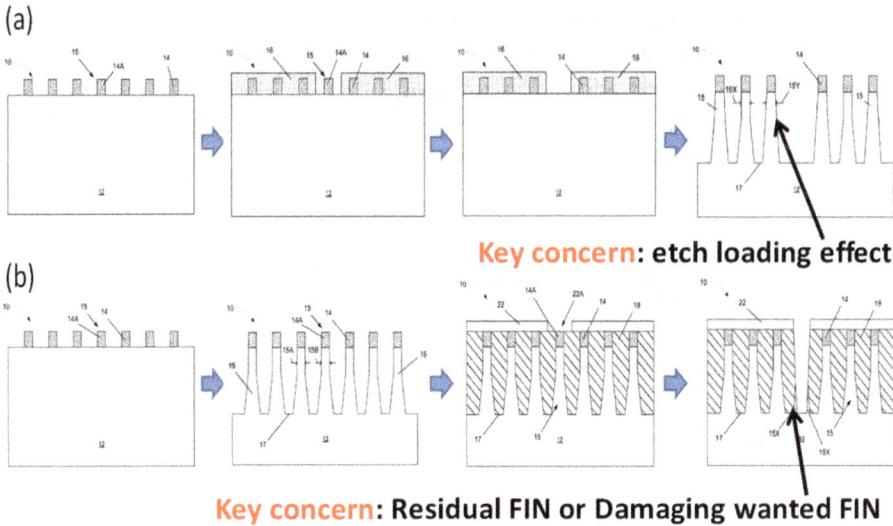

Key concern: etch loading effect

Key concern: Residual FIN or Damaging wanted FIN

Fig. 21. Schematic depicting (a) Fin cut First process (b) Fin cut last process [Source: Reference 80].

Since in the 'fin cut first' scheme, the cut is done on the hardmask level only, there is a much wider process window to prevent the fin spike formation. However, there are also benefits to the 'fin cut last' scheme. Since the final fin etch is performed before the dummy cut, there is no differentia RIE loading effect during the fin formation. This uniform proximity environment during the fin etch improves dimensional and profile uniformity of all the active fins across the devices. In contrast the 'fin cut first' scheme cuts the dummy fin in the initial hardmask before the final fin etch, this results in inevitable RIE loading variation which results in a different fin shape or dimension at the edge of the group of fins forming the active device. Further, the 'fin cut last' scheme allows for dual shallow trench isolation (STI) to get better isolation where the dummy fins are cut. This dual STI addresses another manufacturing challenge related to fin trench depth. If the fin trench depth is too deep, there can be mechanical stability issues for the fins with high aspect ratios, which leads to fin bending. The solution is to have a deeper isolation in the cut area while maintaining shallow isolation on the fin area. This means, that two separate STI processes are required in the fin cut last scheme but this enables deeper fin trench depth across the well isolation. On the other hand, the 'fin cut first' scheme only employs one RIE step which enables only a single isolation depth everywhere.

5.1.3. *Fin etch*

To achieve high drive current, it is better to build taller fins which results in larger effective device width (W_{eff}). But, as mentioned before, these tall fins have to be built with a straight profile and with rounded top corners. If the fin profile is tapered (i.e. if the top CD is smaller than the bottom CD) or has a pointed top corner, the channel is tuned on mostly on the top of the fin. This is due to higher depletion and electric field induction in this region when the gate voltage is applied and makes most of the drive current flow on top of fin. Therefore, the fin profile control has to be optimized to get better performance with taller fin height. Maintaining the fin profile is also very important for downstream processing and device characteristics.

The fins can be etched with and without hard mask during the fin formation. The fin profile is primarily dependent on the etch chemistry and the particular hard mask used on the fin region prior to etching (see Fig. 22). The choice of hard mask material and thickness is critical to obtaining a straight fin profile in narrow fin structures. For example, control of etch byproducts and fin CD loss during the fin etch must be optimized with the choice of hard mask material and thickness[81]. Additionally, an un-optimized etch can create fin side wall damage[82].

Non-optimized   Optimized

Fig. 22. Fin profile comparison between optimized and non-optimized carbon containing etch by-product control during fin etch. In optimized fin etch, SiOC hard mask is fully consumed at the end of fin etch. Therefore, a-C is not exposed to the etch chemical and the carbon containing by-product is not created [Source: Reprinted with permission from Springer Publishing, Reference 81].

5.1.4. *Shallow trench and deep trench isolation*

To maintain sufficient fin height, the fin aspect ratio has to increase as the fin pitch decreases. This also means that both the fin width and the fin pitch decrease simultaneously. High aspect ratio process (HARP) isolation oxide, which is used in planar technology, is no longer useful for shallow trench isolation in aggressively scaled fin pitch. Even though the HARP oxide is filled in the trench by means of a 'deposition-etch-deposition-etch' process, void formation is inevitable across the wafer and presents a major manufacturing challenge. Void formation can lead to active fin height variation, which in turn leads to effective FinFET width variation. There are several techniques to fabricate a void free 'gap fill' in higher aspect ratio tight pitches. A liquid 'flowable chemical vapor deposition' (FCVD) oxide can be a potential replacement for better 'gap fill' for a 5nm space with an aspect ratio of less than 30 and reentrant features[83]. However, densification of the FCVD oxide requires post process annealing of the oxide. The anneal process must be optimized, otherwise, depending on the annealing temperature; the densification process could damage the sidewall of the fins.

5.1.5. *Fin dopant implantation*

One advantage of finFET is the ability to use active fin channels without doping, thereby avoiding the RDF effects described earlier. However, punch through stop (PTS) and well dopants have to be introduced to the body of fin, and these can diffuse into the active fin channel. Also, higher ion implantation energy is required to introduce the dopants throughout the taller active Fin. This can cause more physical or crystalline damage to the active fin area. As the fin width becomes thinner, the damage caused by implants becomes larger. This damage will increase junction leakage and parasitic resistance. Therefore, formation of a 'super steep retrograde well' (SSRW) with minimized physical damage on the active channel is required for aggressively scaled finFET with undoped channels. Fig. 23 depicts the implant induced damage and proposes to use hot implant as the potential solution[84,85].

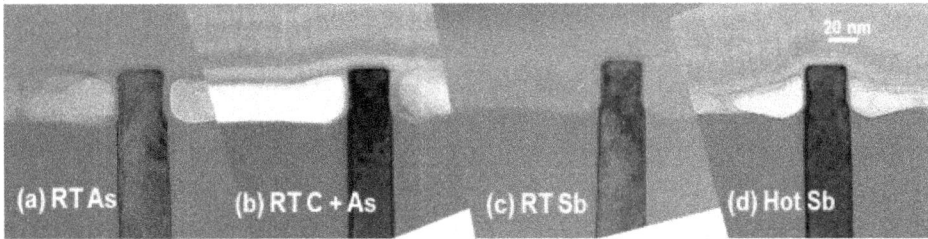

Fig. 23. XTEM micrographs of (a) as-implanted RT arsenic PTS implant, (b) RT arsenic PTS implant co-implanted with carbon, (c) RT antimony implant and (d) hot antimony implant for fin structure wafers [Source: © 2014, IEEE, Reprinted with permission from Reference 84].

5.1.6. *Fin formation of high mobility channel materials*

Fin formation is the primary challenge for alternate channel materials. The channel materials of considerations are (1) $Si_{1-x}Ge_x$ for N & PMOS (2) III-V material such as $In_xGa_{1-x}As$ and $In_xGa_{1-x}Sb$ for N & P respectively (3) combination of $In_xGa_{1-x}As$ and Ge channel for N & P dual channel formation (4) combination of strained Silicon and strained Ge for N & P dual channel formation. Table 4 depicts various combinations for high mobility alternate channel formation along with the respective fin formation approach.

Table 4. Depicts various combinations for high mobility alternate channel fin formation approach.

S/N	Compressively strained PMOS fin	Unstrained or tensile strained NMOS fin	Fin channel formation approach
1	$Si_{1-x}Ge_x$ (x<60%)	Tensile silicon	Blanket growth & Etch (also SRB), Cladding & Condensation, replacement fin
2	$Si_{1-x}Ge_x$ (x>75%)	Tensile Silicon	Blanket growth & Etch (also SRB), Cladding & Condensation, Replacement fin. May not be a practical approach for a fin device due to process related extreme mismatch between Si and high% Ge, in particular thermal mismatch.
3	$Si_{1-x}Ge_x$ (x>75%)	$Si_{1-x}Ge_x$ (x>75%)	Blanket growth & Etch (also SRB), Cladding & Condensation, replacement fin
4	$Si_{1-x}Ge_x$ (x>75%)	$In_{1-x}Ga_xAs$	Blanket growth & Etch, replacement fin
5	$In_{1-x}Ga_xSb$	$In_{1-x}Ga_xAs$	Blanket growth & Etch, replacement fin.

In comparison to Germanium, Silicon is still considered as the best option for NMOS while tensile strained Silicon is considered as the best high electron mobility channel "device material". This is because of the difficulty in forming an electrostatically good and reliable gate with a Germanium channel material. There are various integration approaches to forming alternate channel materials. The simplest approach would be a blanket growth of an alternate channel stack followed by etching to form fins through SADP or SAQP patterning approaches. However, the challenge in such an approach is the formation of a defect free channel material. Alternate channel materials when grown on Silicon form crystalline defects such as dislocations due to the lattice mismatch between the two materials. The next big challenge relates to dual channel formation and is caused by the difficulty of simultaneously etching two different channel materials while maintaining the same fin width for both the materials. Additionally, alternate channel materials are vulnerable to becoming oxidized under silicon optimized thermal processes. The overall process becomes more challenging in dual channel formation where Silicon NMOS and SiGe PMOS are integrated. In particular, shallow trench isolation formation on the alternate channel materials such as SiGe is one such challenging process. Fig. 24 depicts a SiGe fin that was exposed to the conventional Silicon CMOS thermal process. The white spots around the fin are germanium nanocrystals that out-diffused from the SiGe fin into the surrounding oxide. Therefore, low thermal processes are necessary when alternate channel materials are integrated. However, the above mentioned oxidization makes junction and contact formation to fabricate low resistive contacts challenging. One approach to reduce steady state oxidation is to isolate the alternate channel materials from the oxygen ambient by growing silicon nitride around them.

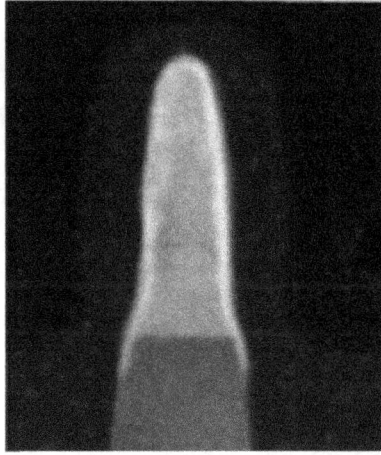

Fig. 24. SEM image of a SiGe fin with un optimized thermal anneal process out diffusing Germanium nanocrystals into the SiO$_2$ matrix.

Another approach would be the 'replacement fin' integration scheme[86,87,88,89,90] where fully formed silicon fins are replaced by selectively etching them away and growing the alternate channel in the opened trench. The material surrounding the trench would be the fin to fin isolation material, perhaps SiO$_2$. This process scheme is more promising as low concentration Ge has shown to provide fewer defective fins while maintaining higher stress along the channel as compared to a blanket growth approach. Other influencing growth factors to reduce defects in this scheme are trench etch and etch chemistries, in-situ interface preparation, growth chemistries, growth time and temperature, and above all the fin width. As the fin width decreases, the pseudo critical thickness increases thus providing longer runs of defect-free fins. Fig. 25(a) shows the schematic for replacement fin formation. Fig. 25(b) to (d) shows high resolution transmission electron microscopy images of Si$_{1-x}$Ge$_x$ (with x>75%) fins grown using the replacement fin approach. Here the epi grown SiGe fin width is less than 10nm and the height is greater than 20nm. The replacement fin integration scheme is a better approach for reducing defects in fins. However, completely eliminating defects all across the wafer for HVM is found to be challenging even for this approach, probably because the process is very sensitive to a variety of growth parameters. Also growth on (100) oriented fins is more defective than growth on (110) oriented fins. Fig. 25(c) shows least defective fins when grown on (100) rotated wafers. It is interesting to see that cross sectional images along the fin on (100) rotated wafers have the smallest number of defects along the fins. Fig. 25(d) shows highly strained fins as evidenced from the high resolution X-ray diffraction 2D space mapping. These results by far are better than any other high concentration Ge fin formation in terms of simplicity in process and channel defects.

(a) Schematic showing Replacement FIN growth process

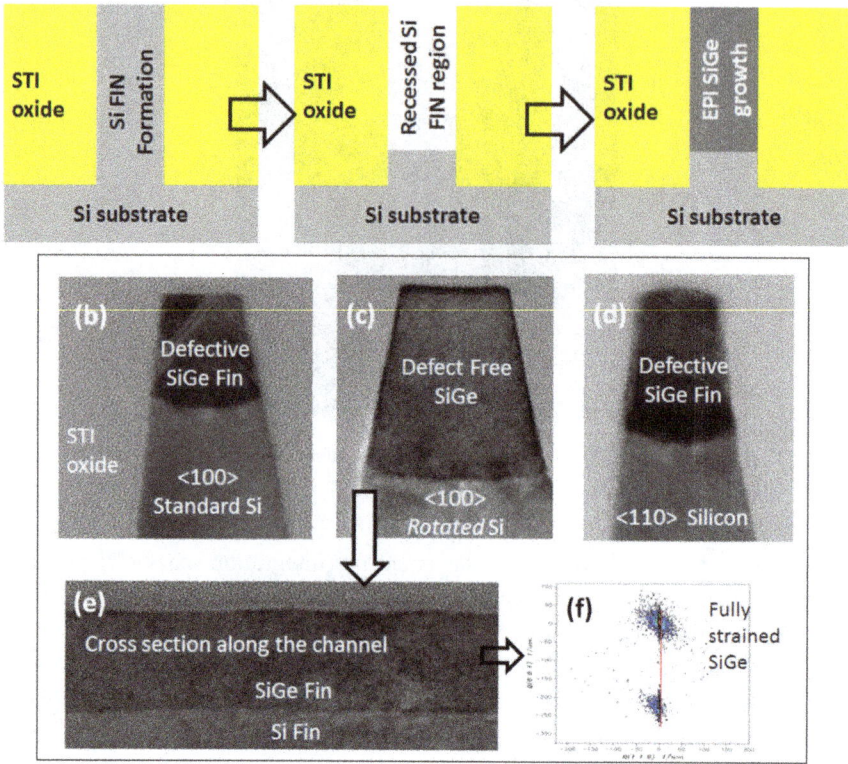

Fig. 25. (a) Schematic showing replacement Fin formation, (b)-(d). High Resolution Transmission Electron Microscopy (HR TEM) images of epitaxially grown sub 10nm SiGe replacement fins grown on various substrate crystal orientations; (b) HRTEM image of defective SiGe fins grown on Silicon fin formed on a <100> standard silicon substrate (c) HRTEM image of least defective SiGe fin grown on Silicon fin formed on a <100> rotated silicon substrate (d) HRTEM image of defective SiGe fins grown on silicon fins formed on a <110> silicon substrate (e) HRTEM image of SiGe fin along the fin channel showing "no" defects in the projected cross sectional area (f) Reciprocal X-Ray diffraction space mapping shows that the SiGe fins grown on the silicon fins formed on a <100> rotated silicon substrate is fully strained.

Another integration option for fabricating SiGe fins is the cladding[91,92] and condensation (see Fig. 26) scheme. In this approach, a very thin layer of SiGe is clad onto the silicon fin and through condensation a high concentration Ge fin can be obtained. The condensation approach oxidizes Silicon in the SiGe and drives germanium into the Silicon fin. Prolonged condensation can generate very high concentration of Ge in the fin. The problem is that the cladding integration scheme is not practical at tight fin pitch because the overall fin width becomes the width of the silicon fin plus the thickness of the cladded region. Fig. 26(a) depicts a schematic of the condensation process, Fig. 26(b) gives high concentration Ge SiGe clad fin and Fig. 26(c) shows the fin after condensation.

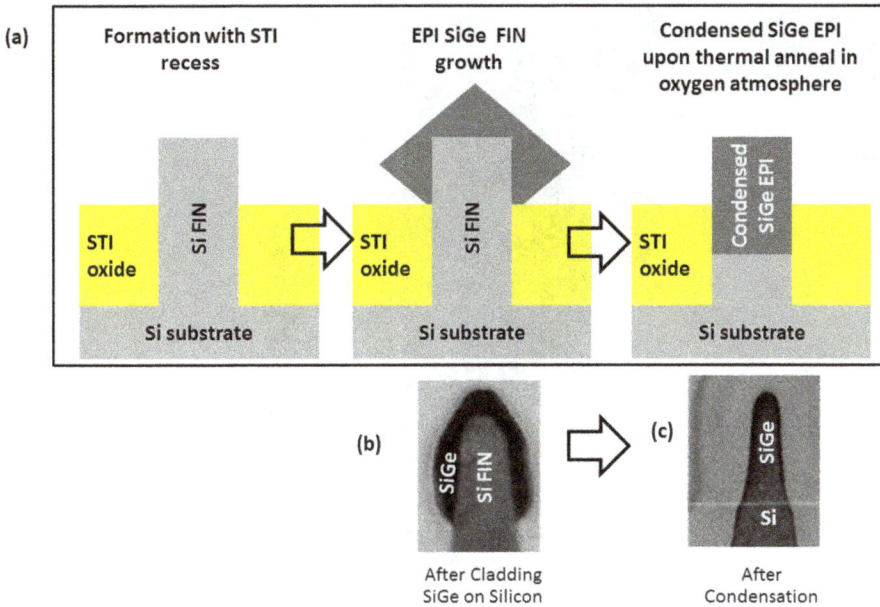

Fig. 26. (a) Schematic showing cladding and condensation alternate channel fin formation (b) HRTEM image of epitaxial grown cladding SiGe on Silicon fin (c) HRTEM image of epitaxial grown SiGe fin formed on Silicon fin after the condensation process.

Monolithic integration of III-V fins on silicon is very challenging due to a much higher lattice mismatch in comparison to SiGe or Ge channel. As in the case of SiGe, these materials can be grown either through a blanket growth and etch process or through a replacement fin approach. The blanket growth process requires a very thick buffer layer to reduce the defect density in the channel. Fig. 27(a) and (b) show sub 10nm GaAs and GaAs/InGaAs heterostructure fins grown directly on silicon using the replacement fin approach. The images show low defect sub 10nm GaAs and InGaAs fins grown on GaAs/Silicon fins; however, they are not devoid of defects at the interface as a cross section of the fins along the channel show. At the wafer-scale, growth of III-V material on a 300mm wafer requires careful process optimization while at the transistor-scale uniform growth is very dependent on pattern loading effects. Any change in fin pitch, silicon fin dimension, or silicon fin surface cleaning prior to the growth impacts the growth uniformity. Regardless, these processes have substantially lower amounts of defects on a fin when grown directly on silicon versus any other current growth scheme. Also, there are a lot of environmental health and safety (EHS) procedures required for the use of group III-V gaseous chemistries during fin growth and subsequent fabrication steps. Optimization of all these approaches is still not mature and is far away from cost effective manufacturing.

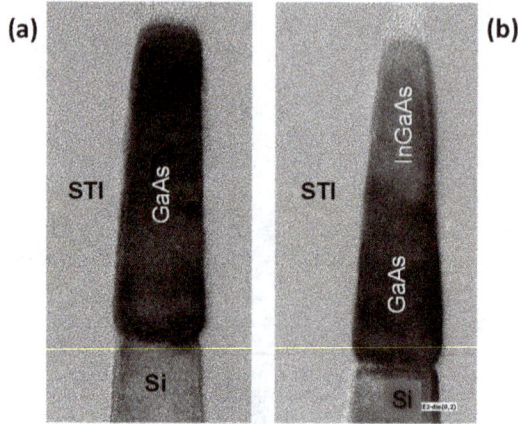

Fig. 27. TEM image of epitaxially grown III-V fins formed directly on silicon fin through replacement fin process scheme. (a) TEM image of sub 10nm GaAs fin grown on Silicon fin (b) TEM image of sub 10nm GaAs/InGaAs fin heterostructures formed on Silicon fin. Both (a) and (b) shows that III-V materials grown directly on Silicon fin can have very low defect density compared to their blanket growth directly on bulk silicon substrate.

5.2. *Challenges in dummy gate and spacer formation*

As gate pitch and gate-length scale, one of the most challenging parts of the integration flow is to maintain the mechanical stability of the gate structure. Fig. 28 shows that gate-bending occurs during the spacer deposition due to the stress introduced by the spacer material[93]. A high aspect ratio gate structure is more susceptible to bending, which poses a significant challenge for finFET scaling beyond the 10nm node.

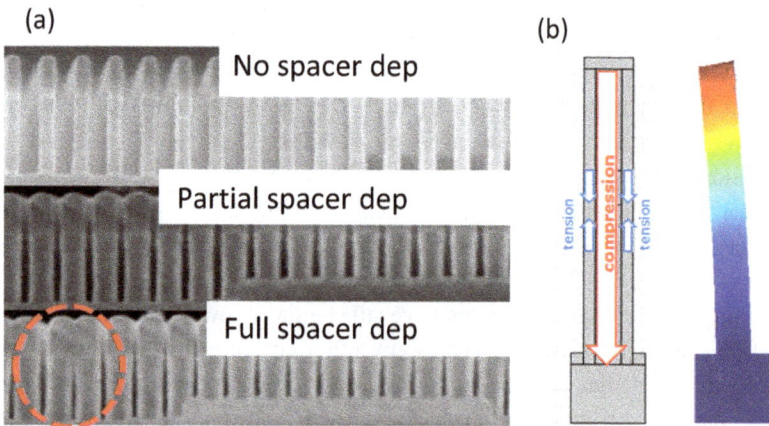

Fig. 28. (a) Gate bending after spacer deposition; (b) Analytical model output of gate buckling behavior.

Another critical challenge associated with the gate pitch scaling is the reduced spacer width. As spacer width becomes smaller, a lot of integration and device challenges arise. One concern is that the contact is getting closer to the gate and the risk of contact-to-gate shorts is getting higher. This concern has been addressed with advances in SAC formation and will be discussed later in session 5.4. Another concern is increasing capacitance between the contact and the gate metal due to spacer thickness reduction. This parasitic capacitance could significantly reduce the AC performance of the device. As a result, a lot of effort has been expended to reduce the relative permittivity of the dielectric material (i.e. the k-value) of the spacer material. To that end, low-k spacer materials like SiBCN and SiOCN have been introduced to replace silicon nitride (SiN). Ultimately, an air-gap spacer may be the final solution to reduce the parasitic capacitance between gate and contact[94]. Fig. 29 shows the air-gap spacer process flow. The most critical and challenging process is to remove the gate cap and spacer after source and drain metallization without damaging the gate stack and fins while leaving an air gap at the spacer region between the gate and the contact.

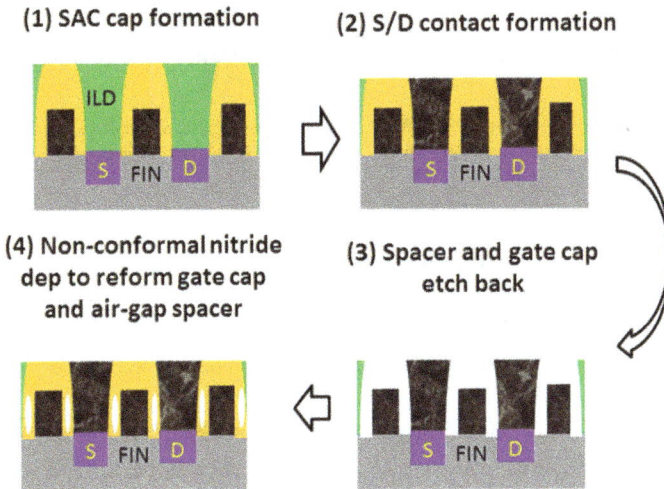

(1) SAC cap formation   **(2) S/D contact formation**

(4) Non-conformal nitride dep to reform gate cap and air-gap spacer   **(3) Spacer and gate cap etch back**

Fig. 29. Proposed process flow of the novel air-spacer SAC transistor (1) After SAC cap formation. (2) S/D contact plug formation. (3) SAC cap and nitride spacer removal. (4) Reform SAC cap and air-gap spacer.

5.3. *Challenges in epitaxial source and drain formation*

As in planar technologies, epi grown epitaxially grown S/D enhance the performance of the finFET by imparting strain to the channel and forming sharp ultra-shallow junctions. Performance improvement depends on the following factors (1) distance between S/D, (2) raised or embedded S/D growth, (3) epi growth profile, (4) S/D etch shape profile, (5) S/D etch depth, (6) volume of Ge material for PMOS, (7) fin pitch, (9) single fin vs multi-gated nested fins, (8) in-situ doping profile, (10) active dopant density, (11) lightly doped profile closer to the junction/channel interface to the heavily doped region near the silicide/contact interface, and (12) for alternate channel materials, the S/D epi should provide compressive strain to the PMOS channel and tensile strain to the NMOS channel.

The distance between the source and drain regions are primarily determined by the S/D junction formation and are thus dependent on the type of epi formation as well: raised S/D and/or embedded S/D greatly influence the device performance. Schematics of raised and embedded S/D are as shown in Fig. 30.

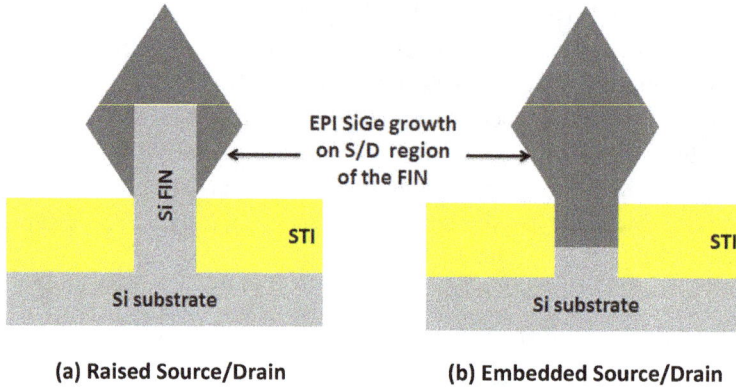

EPI SiGe growth on S/D region of the FIN

Si FIN

STI

Si substrate

STI

Si substrate

(a) Raised Source/Drain   **(b) Embedded Source/Drain**

Fig. 30. Schematic image showing (a) Raised and (b) Embedded Source/drain. In this schematic, the fins run into the page. (a) Raised S/D formation, epi is grown directly on the S/D portion of the fin. (b) Embedded S/D portion of the fin is recessed and epi is grown.

The raised S/D generates biaxial strain while the embedded S/D generates uniaxial strain to the channel wherein the biaxial strain is less than the uniaxial strain. Raised S/D is a natural approach for SOI based finFETs because the embedded S/D retains only very little active silicon in the S/D fin region for epi S/D growth. Also, the S/D etch can reduce the existing strain in the fins.

Raised S/D epi grown on a (100)/<110> channel can create diamond shaped SiGe growth on the fin S/D. With this, two types of S/D epi can be formed: merged and un-merged epi (see Fig. 31), but both types of have device challenges. For multigate transistors with an aggressive fin pitch, merged epi is a natural process due to space constraints. However, it generates epi defects when the competing crystal planes from adjacent fins merge. These defects can sometimes propagate to the fin region as well. Further, defect formation during the epi merge reduces the strain imparted onto the channel. Merged S/D epi is also a challenge for creating better contacts, such as wrap around contacts (WAC) that could reduce the contact resistance. On the other hand, the challenge with un-merged S/D epi is that, due to the low epi volume, the amount of strain generated will be far less than that of any non-defective merged epi. In addition to the reduced strain, the low epi volume also results in low dopant density, thus increasing S/D resistance. However, unmerged epi can support WAC that could result in reduced resistance.

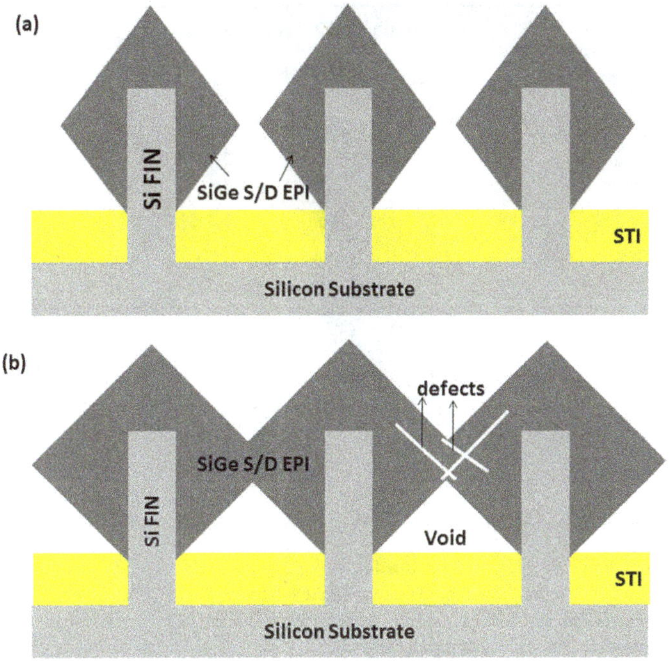

Fig. 31. Schematic image showing two types of S/D epi formation (a) Un-merged diamond shaped epi formation. Due to the low epi volume, impact of strain on the channel due to unmerged epi is limited and (b) Merged epi S/D formation; Merge can generate defects at the merging crystal planes and propagate it to the channel.

Embedded S/D is a better performance booster for bulk fins; however, it does not provide the equivalent performance boost as in the planar transistors. The 'sigma shape' that generates the maximum strain in planar FETs gives the least amount of strain in a finFET. The maximum strain is obtained from a 'U shaped' etch profile as compared to other profiles such as rectangular and sigma shapes (see Fig. 32 and Fig. 33)[95]. This essentially means that for the same amount of Ge concentration and the volume, the channel cannot be stressed to the same level as in a planar transistor. However, as in a planar transistor, the stress is proportional to the etch depth (see Fig. 34) which is proportional to the epi volume. This dependency is most pronounced in SOI vs bulk S/D epi profiles[75]. SOI devices have low substrate leakage, but due to their inability to accommodate a large embedded epi volume to provide compressive stress to the channel, they have lower performance in comparison to the bulk finFETs.

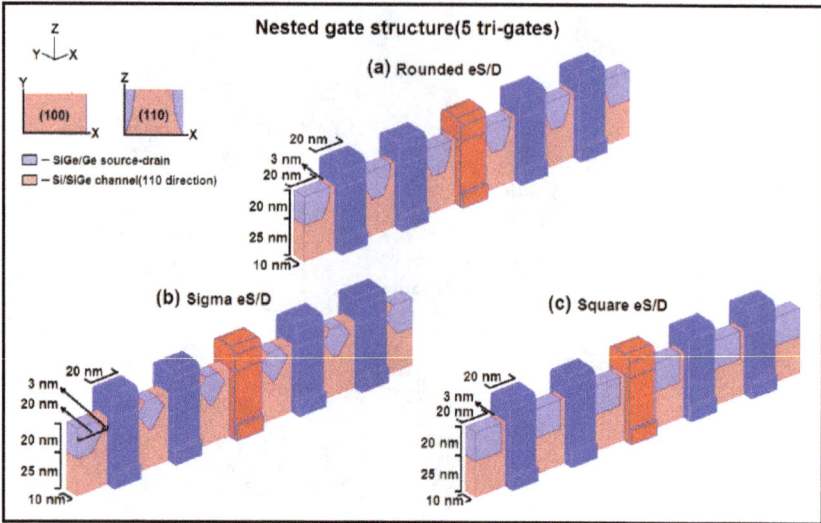

Fig. 32(a)-(c). Schematic of simulation structures for evaluating the impact of S/D shape and channel content to the average channel stress keeping the same lattice mismatch between channel and the S/D regions. (a) Epi S/D in a round shaped trench (b) Epi S/D in a sigma shaped trench (c) Epi S/D in a square shaped trench [Source: Reprinted with permission from S. S. Mujumdar, Reference 95].

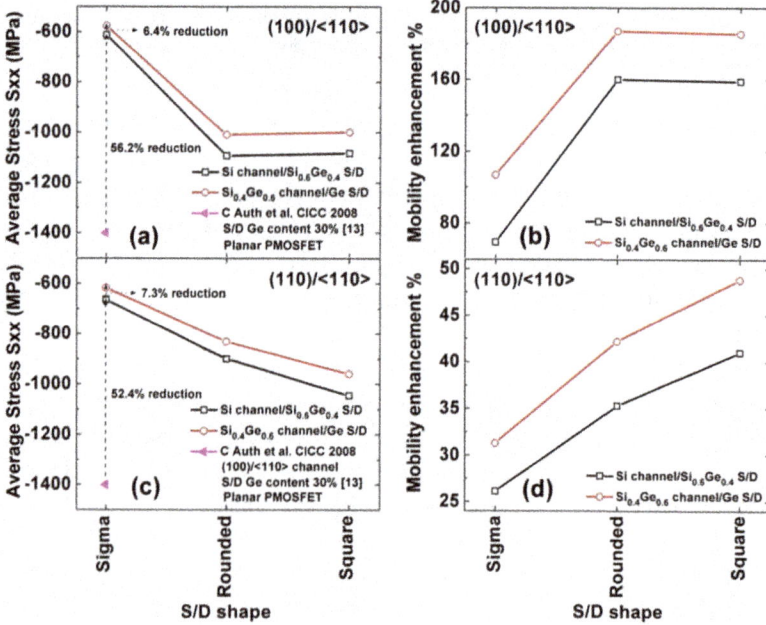

Fig. 33. Average sidewall stress and mobility plots for various S/D etch and corresponding crystal orientation in accordance to Fig. 32. Fig. 33(a) and (c) depicts sidewall stress vs S/D shape for (100)/<110> and (110)/<110> channel stress; Fig. 33(b) and (d) gives corresponding mobility enhancement plots for (100)/<110> and (110/<110> channel stress [Source: Reprinted with permission from S. S. Mujumdar, Reference 95].

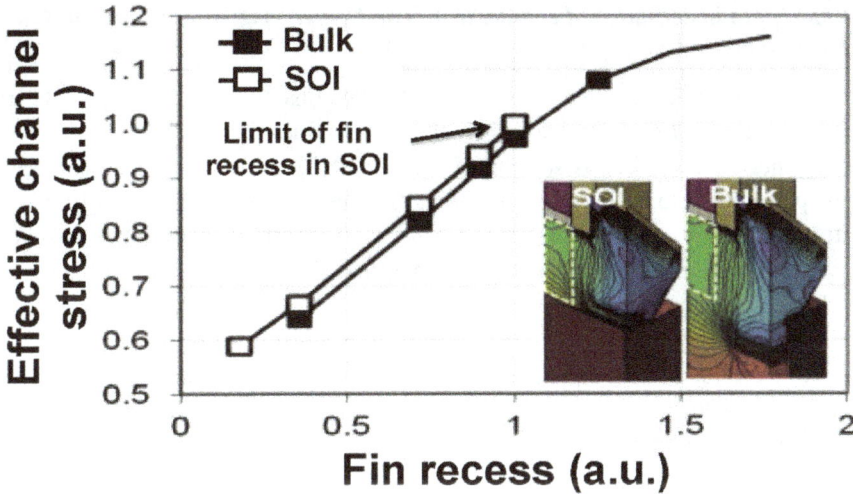

Fig. 34. Effective channel stress as a function of fin recess for a short channel PFET. Deeper fin recess in bulk allows larger stress at the channel [Source: © 2014, IEEE, Reprinted with permission from Reference 75].

It should also be noted that the embedded strain is dependent on the channel orientation of the fin: (100)/<110> provides higher strain boost as compared to (110)/<110> channels. As in raised S/D, in embedded S/D the epi formation can also be merged or unmerged. As with raised S/D the merged epi generates defects into the structure and also at the interface where adjacent epi merges, again reducing the strain in the channel. Also, a diamond shaped epi could leave behind voids under the epi merge region which is an added reliability concern. As the fin pitch reduces, another potential risk is that the larger diamond shaped epi can short to a neighboring device's epi. Yet a smaller epi will increase the S/D contact resistance in both N & PMOS because of the reduced dopant concentration. Smaller epi can also reduce the compressive stress in PMOS thus further reducing the total performance. NMOS epi growth such as Phosphorous doped Silicon also has similar epi growth challenges. epi defects increase as the P doping increases in the epi.

For alternate channel materials such as SiGe, S/D formation is a major challenge. To obtain a defect free alternate channel, it is preferable to grow them strained: compressive for PMOS and tensile for NMOS. For example, channel SiGe grown on Silicon is compressively strained and Silicon grown on SiGe will be tensile strained. These channels, if strained, will get relaxed during S/D etch. For a pure Germanium-based PMOS channel the compressive strain could be provided by alternate S/D materials such as Germanium-Tin (GeSn) which has higher lattice constant. However, higher Sn concentration in Ge is a challenge because of the solid solubility limit which limits the scaling of compressive strain in germanium based devices. For a pure Germanium-based NMOS channel tensile strain could be provided by SiGe. Carbon doped S/D such as SiGe:C could provide tensile strain, however, Carbon, being a smaller atom, is not stable inside the lattice. Tensile strain could also be engineered on InGaAs based III-V NMOS channels.

5.4. *Challenges in self-aligned contact (SAC) and replacement metal gate (RMG) formation*

In addition to the challenges due to the introduction of fins, the gate pitch and gate length scaling also create a lot of integration challenges. As shown in Fig. 35 in older planar technology nodes, gate pitch is so relaxed such that S/D contacts and gate contacts can easily be placed next to each other without causing any shorting risk (see Fig. 35(a)). As the gate pitch scales, there's no room to put gate contacts next to S/D contacts, and gatecontacts have been pushed away from the active region and are only placed on the STI region. In addition, at tight gate pitch, even forming S/D contact without shorting to gate metal becomes very challenging. The idea of self-aligned contacts (SAC) has been introduced to mitigate the issue of S/D contact to gate shorts. As shown in Fig. 35(b), the gate metal is fully encapsulated by a dielectric spacer and gate cap, which protects the gate from shorting to the S/D contact.

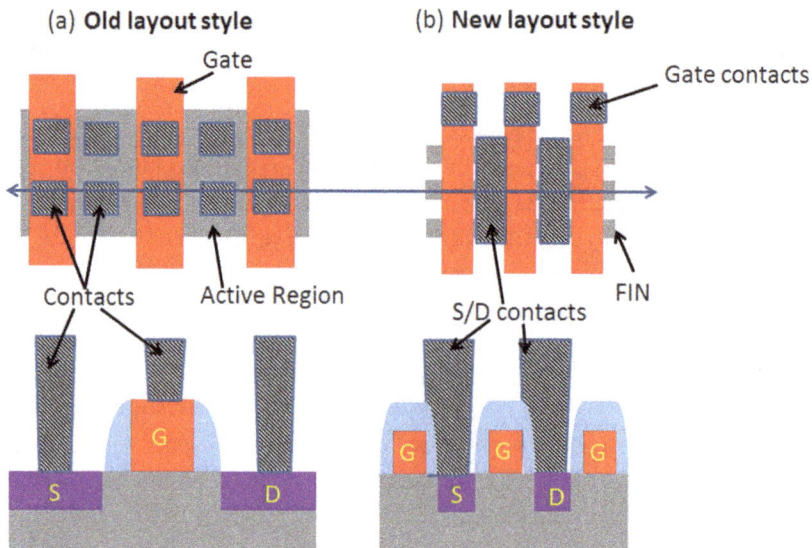

Fig. 35. (a) and (b) Layout style for older technology node and the current non-planar technology node respectively as gate pitch scales.

Forming SAC for gate first technology can be straightforward[96]. It can be conveniently achieved by depositing a dielectric hard mask (HM) layer during the gate patterning process and, after spacer formation; the gate metal is fully encapsulated. To further improve the SAC etch selectivity, an additional etch stop liner can be deposited before ILD dielectric fill (see Fig. 36(a)). Fig. 36(b) shows the TEM image of SAC formed with gate first patterning process. The full metal gate (FMG) is fully encapsulated by a nitride spacer and HM as well as an HfO_2 etch stop liner which has been used to further enhance the SAC etch selectivity.

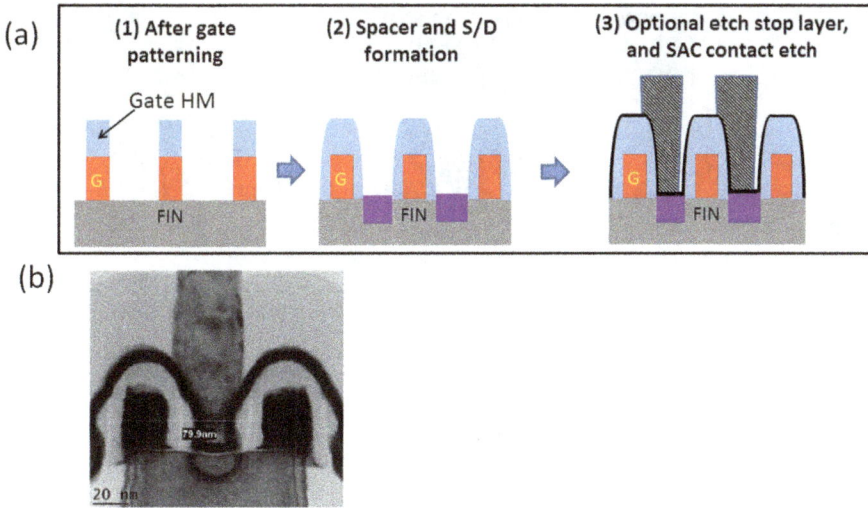

Fig. 36. (a) Gate first self-aligned contact (SAC) formation flow; (b) TEM x-section image of FMG with SAC contacts at 80nm CPP [Source: © 2011, IEEE, Reprinted with permission from S.-C. Seo *et al.*, Reference 96].

Although the gate first integration flow favors SAC formation, it is not adopted by mainstream technology nodes, because of following two reasons: (1) the gate etch process is very challenging with FMG materials, especially since NFET and PFET have different metal stacks with different thicknesses. Also the gate etch needs to remove the metal portion wrapping around the fins, (2) geometric effects in the etch play a significant role as gate length scales; e.g. the impact of plasma damage during gate etch or wet process damage during post gate etch clean at edge of the channel become increasingly significant as the gate length becomes shorter.

The replacement metal gate (RMG) process flow mitigates the above issues. However, forming the SAC contacts for a RMG integration flow can be very challenging. As shown in Fig. 37, in a RMG process flow, to fully encapsulate gate metal with dielectric, additional processes like RMG gate recess and SAC cap formation need to be performed. Particularly, the RMG gate-recess process can be very difficult. As illustrated in Fig. 38, the RMG is usually formed by depositions of multiple layers such as high-k gate dielectric, work function metal (WFM) and low resistance metal (e.g. W) depositions. As a result, the top surface of the gate after chemical mechanical polish (CMP) has a very complex composition. It is very difficult to find a proper etch process to etch all the different materials uniformly across the wafer and on different gate-lengths. One method to mitigate this issue is to implement WFM chamfering[97,98]. As shown in Fig. 39(a), the WFM chamfering process avoids recessing complex materials by etching only one material at a time. This is enabled by first depositing a sacrificial material, then etching back the sacrificial material, and finally removing the exposed WFM1 material. To complete the process, the sacrificial material is also removed. By repeating the processes for WFM2, WFM3…, the RMG can be recessed in a more controlled manner (see Fig. 39(b)).

(1) After dummy gate CMP

(2) Dummy gate removal

(3) RMG gate formation

(6) SAC contact formation

(5) SAC cap formation

(4) RMG gate recess

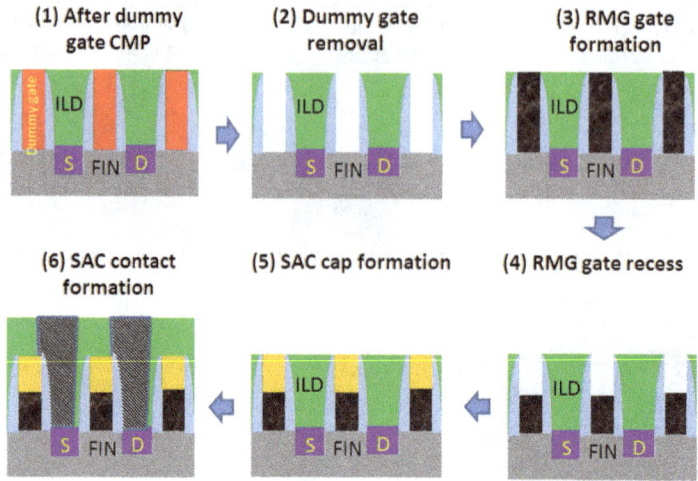

Fig. 37. SAC contact formation with RMG flow.

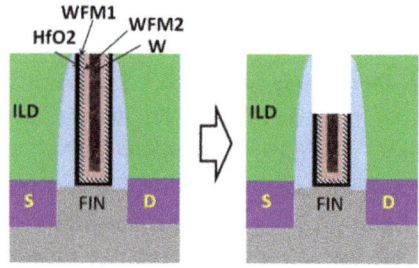

Fig. 38. Illustration of RMG gate recess process.

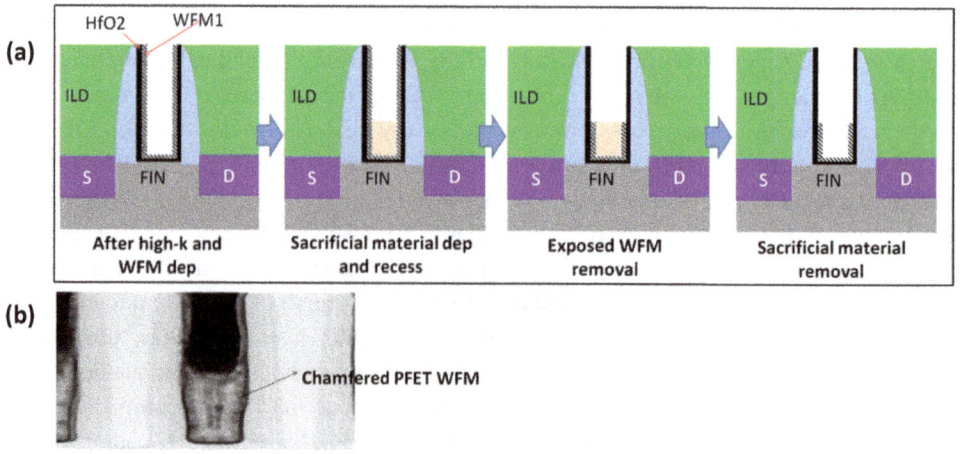

(a)

After high-k and WFM dep

Sacrificial material dep and recess

Exposed WFM removal

Sacrificial material removal

(b)

Chamfered PFET WFM

Fig. 39. (a) Schematic of work function metal (WFM) chamfering process flow (b) TEM of a gate RMG process with PFET WFM chamfering process.

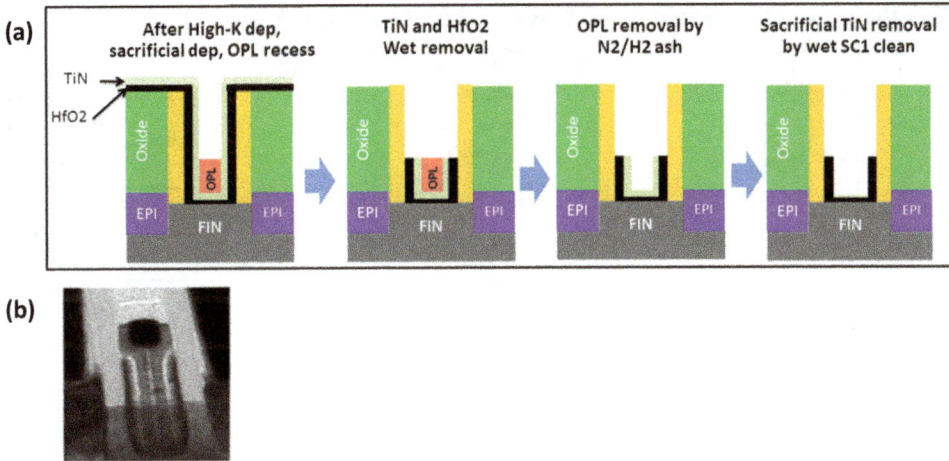

Fig. 40. (a) Illustration of high-K chamfer ing process (2) demonstration of controllable RMG gate process achieved @ Lg = 15nm using high-K chamfer process.

However, even with the WFM chamfering process being used, the RMG gate-recess becomes extremely difficult when the gate dimension is scaled to less than 17nm. This is because after the high-k and WFM deposition, the remaining gate opening for the sacrificial material is very small (can be only ~5nm). Filling and recessing such a small volume of sacrificial material becomes very hard and can have large process variation. At these dimensions, high-k chamfer as shown in Fig. 40 can be introduced to widen the top gate CD and improve the sacrificial material fill and recess for WFM chamfering[99]. The high-k chamfer process is very similar to WFM chamfering. The key difference is the additional TiN layer to protect the high-k material before the sacrificial material such as organic planarization layer (OPL) deposition. The sacrificial material removal process, such as oxygen ashing, can impact the gate stack quality, leading to Tinv increase or high-k damage.

5.5. *Challenges in diffusion break formation: Single diffusion break vs double diffusion break*

Another important feature that impacts scaling is the diffusion break. Diffusion break refers to the space separation between two active device regions. Historically, the 'diffusion' area has been the silicon (non-isolation) regions, so now it represents the fin regions. As shown in Fig. 41, the diffusion break can be designed as 'single diffusion break', where only one dummy gate separates the two active regions, or 'double diffusion break', where two dummy gates separate the two active regions. Of course, the goal of scaling would be to minimize the number of dummy gates as these only consume space without adding functional value.

Single diffusion break

Double diffusion break

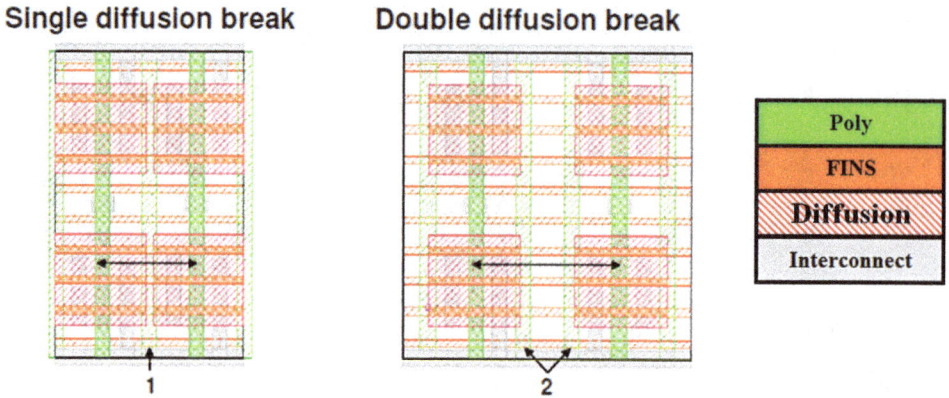

Poly	
FINS	
Diffusion	
Interconnect	

Fig. 41. Layout of single diffusion break and double diffusion break.

In advanced technology nodes, the double diffusion break is most widely used in the industry and the first generation finFET already utilized this design[97]. As shown in Fig. 42, two dummy gates are used to tuck-under the fin-end. This provides a process window to tolerate the STI CD variation and gate to STI misplacement. On the other hand, if single diffusion break is used to gain higher active transistor density, Fig. 43, the STI dimension needs to be very narrow so that the fin-ends of both adjoining finFET can tuck under one narrow dummy gate, even under worst case misalignment. Patterning and etching such a narrow STI is very difficult and narrow STI CD can lead to high device leakage due to poor isolation.

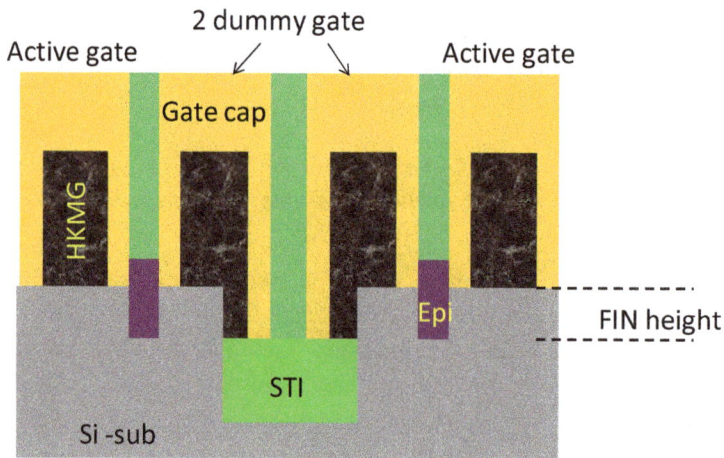

Fig. 42. Schematic illustration of double diffusion break. Two dummy gates are used to tuck the fin ends. It provides good process window to tolerate STI CD variation and gate misplacement.

Single diffusion break

The tiny dummy gate has to tuck both FIN ends

~10-20nm wide deep-trench isolation

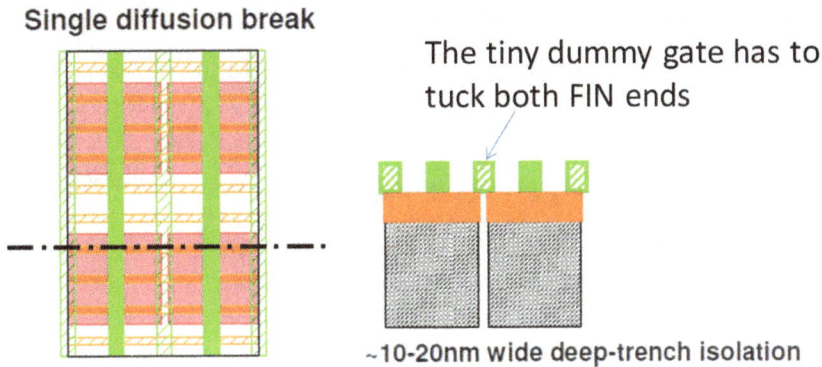

Fig. 43. Process window for single diffusion break is very small because one dummy gate needs to tuck both fin ends. This requires very narrow STI width and very accurate gate placement during patterning process.

Although a single diffusion break is very challenging to manufacture, the area scaling benefit is significant. By reducing the dummy gate area consumption, the overall logic cell area can be reduced by up to 20%, providing a huge scaling benefit. To overcome the gate placement issue on the fin ends, self-aligned isolation techniques can be used[100]. These cut the active region later after the dummy gate removal process and either fully fill the gate with dielectric or partially fill the top region with high-k metal gate (see Fig. 44).

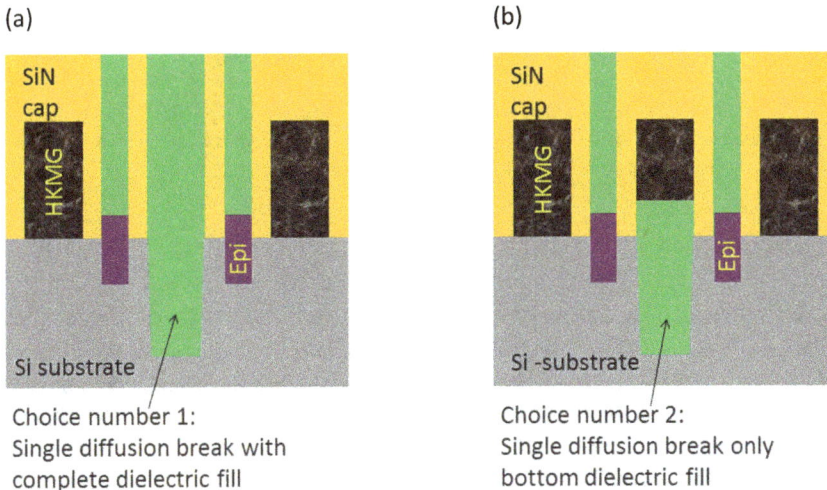

(a)

SiN cap

HKMG

Epi

Si substrate

Choice number 1:
Single diffusion break with complete dielectric fill

(b)

SiN cap

HKMG

Epi

Si -substrate

Choice number 2:
Single diffusion break only bottom dielectric fill

Fig. 44. Single diffusion break self-aligned to dummy gate. (a) Dummy gate full filled by dielectric; (b) Dummy gate is partially filled with high-k metal gate (HKMG).

5.6. *Challenges in source drain contact formation*

Further improvement in device performance is limited by the increasing contribution of parasitic series resistance to device resistance. The resistance of a device can be expressed as

$$R_{dev} = R_{ch} + R_{para} \qquad (4)$$

where R_{Dev} is the device on-resistance, R_{Ch} is the channel resistance and R_{para} is the parasitic series resistance. R_{para} is modeled with these five resistance components: (1) overlap resistance R_{ov}, (2) extension resistance R_{ext}, (3) source/drain resistance (R_{sd}), (4) contact resistance for silicon-to-silicide $R_{con\text{-}silicide}$, and (5) contact resistance for silicide-to-contact $R_{silicide\text{-}con}$. Among these resistance components, R_{ext}, $R_{con\text{-}silicide}$, and $R_{silicide\text{-}con}$ are projected to contribute equally at the 5nm technology node (N5) as shown in Fig. 45. This leads to a R_{para} limited scaling regime in leading edge technologies.

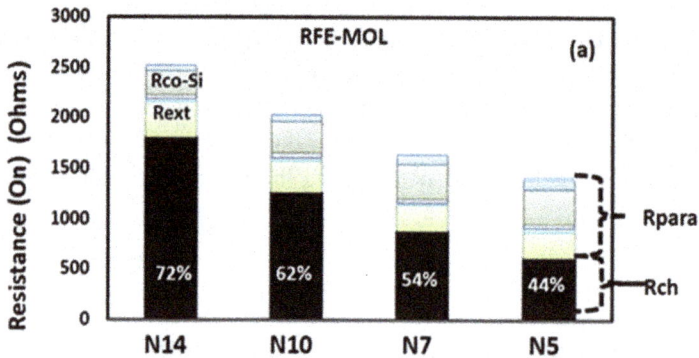

Fig. 45. FEOL + MOL resistance breakdown between on-state channel (R_{ch}) and parasitic series resistance (R_{para}) for leading technology nodes. N14 to N5 represents respective technology nodes [Source: © 2015, IEEE, Reprinted with permission from A. V. Y. Thean *et al.*, Reference 162].

With the introduction of the finFET architecture, the key challenge for R_{ext} reduction is the fabrication of highly doped conformal and damage-free fins to form the S/D extension regions. The sidewalls of a fin can be doped using conventional beamline implantation with the implant angle restricted to 10 degrees to avoid shadowing in aggressively scaled fin pitch (see Fig. 46(a))[101]. However, the high implantation angle increases backside scattering and leads to an exponential loss in implanted dose (see Fig. 46(b))[102]. Additionally, ion implantation can lead to full amorphization of the fin and problematic recrystallization, resulting in defect formation and poor activation of the dopants[85]. Several alternative doping techniques have been investigated to overcome this issue: hot implantation[101], plasma doping[103], vapor phase deposition[104], and solution-based monolayer doping[105,106] are examples. These alternative doping techniques will become even more relevant for subsequent technologies such as nanowires and vertical FETs.

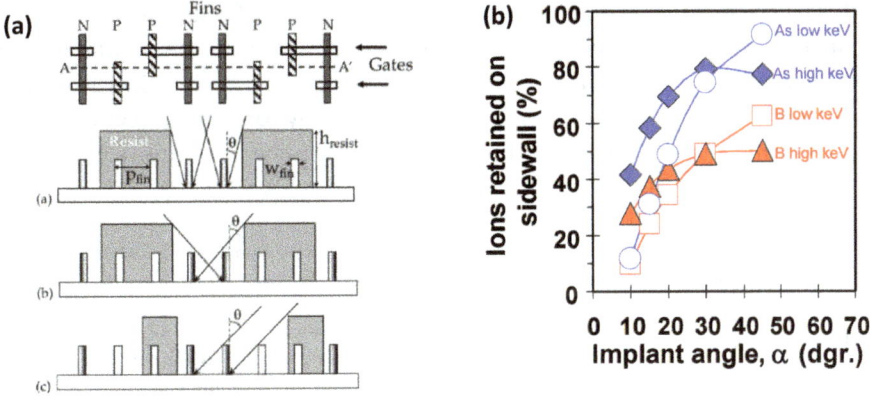

Fig. 46. Plan view and cross-sectional schematics for a dense SRAM illustrating photoresist pattern and limitation on implant tilt angle for three approaches: (a) Standard, (b) 2 × 1 and (c) 1 × 1, (b) Simulated B and As sidewall dose retention versus implant angle in 10nm wide fin [Source: Reprinted with permission from Reference 101].

To optimize the $R_{silicide-con}$ component, one can look to the Equation 5 describing this component as

$$R_{silicide-con} = \frac{\sqrt{\rho_{co} R_{sheet}}}{W_c \times tan\left(\frac{L_c}{L_T}\right)} \tag{5}$$

where ρ_{co} is the contact resistivity, R_{Sheet} is the S/D sheet resistance, W_C and L_C are the width and length of the contact hole respectively, and L_T is the transfer length. To lower $R_{silicide-con}$, we can reduce ρ_{co} according to Equation 6. This is accomplished either by reducing the metal Schottky-Barrier Height and/or increasing the semiconductor active doping concentration (N_D).

$$\rho_{co} \propto e^{\left[\frac{4\pi\sqrt{\varepsilon_s m*}}{h}\left[\frac{\Phi_B}{\sqrt{N_D}}\right]\right]} \tag{6}$$

where ε_s is the permittivity of the semiconductor, $m*$ is the effective mass of the semiconductor, h is the Planck's constant, Φ_B is the Schottky barrier height (SBH) between the diffusion region and metal contact layer, and N_D is the active dopant concentration in the semiconductor, and the mechanism for current flow through these ohmic junctions is assumed to be tunneling. From Equation 6 it is obvious that ρ_{co} has an exponential dependence on Φ_B and N_D. Hence, there is strong motivation to develop solutions to reduce SBH and/or increase N_D to boost MOSFET drive current performance. For this purpose, metal silicides have become an important aspect of leading edge technologies as the formation of metal silicides results in an intimate metal-semiconductor interface thru the reaction of the deposited metal and underlying semiconductor. This eliminates the interfacial oxide layer between the metal and semiconductor and thus reduces $R_{silicide-con}$.

To further reduce $R_{silicide-con}$ there has been a significant amount of research (see Table 5 for metal silicide to silicon CMOS)[107] into aligning the metal silicide SBH with either the conduction or valence band of all technologically important semiconductors (i.e. Si, SiGe, Ge, etc.). This band alignment reduces the thermionic energy barrier that electrons or holes have to overcome for current conduction from metal to semiconductor, which reduces $R_{silicide-con}$ further.

Table 5. Key parameters of technologically important silicide materials for Si CMOS.

Silicide	Thin film resistivity ($\mu\Omega$-cm)	Formation Temp. (°C)	Si stable up to (°C)	nm Si consumed per nm metal	nm of resulting silicide per nm metal	Barrier Height to n-Si (eV)
CoSi₂	14-20	600-800	~950	3.64	3.52	0.65
TiSi₂ (C49)	60-70	500-700	-	2.27	2.51	0.0
TiSi₂ (C54)	13-16	700-900	~900	2.27	2.51	0.58
NiSi	14-20	350-600	~650	1.83	2.34	0.60
NiPtSi	14-30	350-600	~750	1.83	2.34	0.65
PtSi	28-35	250-400	~850	1.12	1.97	0.84

A dual metal silicide integration scheme has been proposed[108,109] to minimize the thermionic energy barrier for both electrons and holes to achieve an ultra-low $R_{silicide-con}$ for N-FETs and P-FETs, respectively. For example, the lanthanide series of materials such as erbium, ytterbium with low work functions would be ideal for forming a low electron Schottky-Barrier Height for N-FETs[110,111]. Conversely, materials such as platinum and palladium with high work functions are ideal for forming a low hole Schottky-Barrier Height for P-FETs[112]. However, low workfunction materials are highly reactive and oxidize upon exposure to ambient air, which inhibits metal silicide formation[113]. Hence a full in-situ process, from metal deposition to metal anneals, will be required for low workfunction materials. To overcome this issue, additive doping into NiSi to alter its workfunction has been developed[114,115,116]. This doping is achieved by adding small amounts of low workfunction materials into Ni prior to metal silicidation, which forms an alloyed Ni-based metal silicide, which in turn has a modified workfunction and is resistant to oxidation. Furthermore, the dual metal silicide approach for CMOS also requires the development of new processes such as new selective wet etch chemistries and thermal processing schemes. There is also a need for additional lithography steps, which add complexity and cost to a dual metal silicide approach. In addition, a metal-semiconductor interface is pinned near the charge neutrality level, leaving only a fixed amount of SBH tuning available for electrons and holes[117]. This effect is also known as 'Fermi Level Pinning', which is due to the electrons from the metal side charging the interface states within the semiconductor bandgap (intrinsic states) to form what is known as metal induced gap states (MIGS) (see Fig. 47). These states are found to be mostly donor-like when near the valence band (E_V) and acceptor-like when near the conduction band (E_C). The energy level in the band gap at which the dominant character of these interface stages changes from donor-like to acceptor-like is termed as the 'charge neutrality level' (CNL) for MIGS and 'trap neutrality level' (TNL) for defect states.

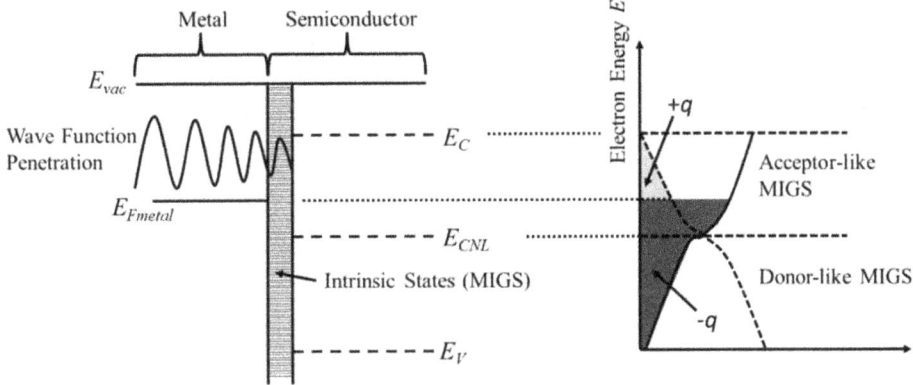

Fig. 47. Fermi level pinning occurs at the metal-semiconductor interface due to the penetration of the metal wave function from the metal to semiconductor side leading to metal induced gap states as shown in the energy band diagram. This could result in the interface being more acceptor like towards E_C and donor like towards EV depending on the alignment of E_{Fmetal} and E_{CNL}.

To eliminate Fermi Level Pinning, the insertion of an ultrathin dielectric between the metal and semiconductor has been proposed and demonstrated and is known as metal-insulator-semiconductor (MIS) or tunneling contacts[118,119,120]. Two hypotheses based on the origin of the Fermi Level Pinning have been proposed[121]. The metal-induced-gap state model states that when a metal and semiconductor are in contact, the metal electron wave function penetrates into the semiconductor bandgap. This charges the semiconductor's intrinsic interface states and subsequently moves the Fermi level at the interface towards the charge neutrality level of the gap states. By inserting a thin insulator at the metal-semiconductor interface, the metal wave function is attenuated in the dielectric and does not penetrate into the semiconductor. This reduces the charges available to drive the Fermi level towards the charge neutrality level. On the other hand, the bond polarization model suggests that the interaction of the metal and semiconductor wave functions forms an interface having both metal and semiconductor-like electronic states. This results in the formation of an interface dipole that pins the Fermi level. According to this model, when a thin insulator is inserted, an additional dipole is formed between the insulator and the semiconductor native oxide. This induces a Schottky Barrier Height shift to offset the pinned electron barrier height. Reported data in the literature suggest that the concept of MIS or tunneling contacts could potentially help to reduce $R_{silicide-con}$ further but it is not without its integration challenges such as the thermal stability of the dielectric[122] and/or applicability of this concept to highly doped semiconductors[123,124,125,126].

As discussed, ρ_{co} also has an exponential dependence on the active doping concentration N_D and a significant amount of work has been devoted into increasing the active doping concentration at the metal-semiconductor interface. The approaches to increasing N_D include: dopant segregation[127,128], novel dopants[129,130,131,132], co-doping[133,134], and ultra-fast high temperature activation[135,136,137]. The idea behind dopant segregation is that a heavily doped silicon layer formed at the silicide/silicon interface causes a strong conduction/valence band-bending near the interface, leading to an effective lowering of the SBHs. Two different schemes have been studied for the introduction of dopants to the silicide/silicon interface: silicidation-induced dopant segregation and silicide as a diffusion source by implantation into silicide followed by a drive-in anneal. The general challenges with all these three dopant-based approaches include the complexity/cost of additional lithography layers for N- and P-FETs and the out-diffusion of these dopants at the S/D regions into the channel of aggressively scaled devices.

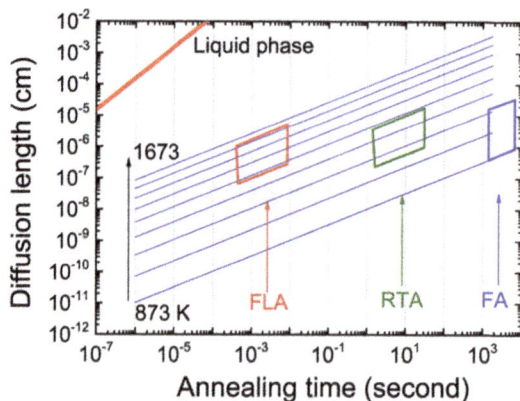

Fig. 48. The three boxes indicate the working regime of different thermal process in solid phase: low temperature furnace annealing (FA) for more than 1000 second, rapid thermal annealing (RTA) for seconds and flash lamp annealing (FLA) for milliseconds at high temperature. The diffusion length of selenium in liquid phase is also shown for comparison [Source: Adapted with permission from Reference 138].

To reduce/eliminate the out-diffusion of dopants in these approaches (i.e. dopant segregation, novel dopants and co-doping) while maintaining or increasing active doping concentration, ultra-fast thermal anneals have been explored extensively. Fig. 48 shows the diffusion length of a dopant, selenium in this example,[138] which exhibits a strong dependence on temperature and time. Hence with careful design and optimization of the thermal processing temperature and time, hyper-doped S/D regions can be achieved with N_D levels beyond the solid solubility limits. The challenge here is to ensure that these hyper-doped and metastable regions remain active through the downstream processing steps.

In the next section, we highlight the key challenges of S/D contact engineering for germanium and compound semiconductors. For germanium, P-type contacts are ease to fabricate due to Fermi Level Pinning, which is exploited for low hole barrier heights to germanium P-FETs. Fig. 49(a) clearly shows that metals with approximately 1.5 eV difference in workfunction are all pinned towards the valence band of germanium[139,140] This indicates that very low-hole barrier heights can be achieved for most metals on germanium for optimal P-FET operation[141,142,143]. The challenge for germanium FETs is in the formation of N-type contacts where Fermi Level Pinning becomes an issue for germanium N-FETs. This can be understood by examining Fig. 49(a) again, good N-type contacts exhibit low electron barrier height to the conduction band (CB), which is impossible on germanium due to Fermi Level Pinning. Furthermore, solid solubility of most N-type dopants[144] in germanium is $< 1 \times 10^{20}$ cm^{-3}. These two issues lead to high electron barrier heights and low active doping concentration for germanium N-FETs. To overcome these issues, similar approaches to Silicon have been proposed for Germanium such as: metal-insulator-semiconductor (tunneling) contacts, co-doping, and alternative dopants and/or ultra-fast anneals. Another challenge unique to germanium semiconductors for MOSFET application is the poor thermal stability of metal Germanides as S/D metal contacts, which is critical for CMOS process integration. It has been shown that metal germanides agglomerate at temperatures as low as 500 °C, which is undesirable for device applications. The use of additive impurities[145,146] and interlayers[147,148] in metal germanides has been demonstrated to enhance thermal stability.

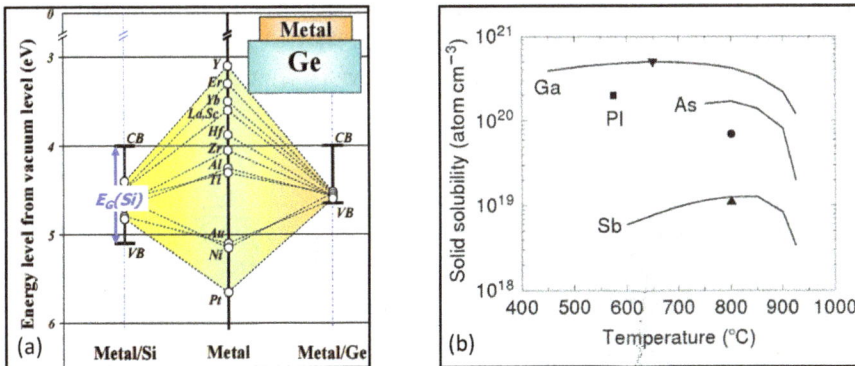

Fig. 49. (a) At the metal/Ge contact, band alignment is not determined by metal work function. Fermi level of metal is strongly pinned to the valence band edge of Ge, hence Schottky characteristics is observed on n-type Ge and ohmic ones on p-type Ge irrespective of the metal [Source: Reprinted with permission from Reference 140]; (b) Solid solubility limits of various dopants in germanium [Source: Reprinted with permission from Reference 144].

For compound semiconductors, specifically Indium gallium arsenide (InGaAs), alloys are being developed to replace Si for N-FET devices. In the case of InGaAs, N-type contacts are easier to fabricate than in germanium due to favorable Fermi Level Pinning characteristics[149]. It has been demonstrated that all metals are pinned toward the conduction band of InGaAs. This leads to low electron barriers for N-type contacts in InGaAs FETs[150,151,152]. However, N-type dopants in InGaAs come from either group IV or group

VI elements. Group IV dopants are referred to as 'amphoteric'. It is assumed in literature that amphoteric dopants such as C, Si, Ge, and Sn are limited in activation due to their propensity to exist in both a donor and acceptor configurations, thereby causing self-compensation[153,154]. Amphoteric compensation may be a downside to the use of group IV dopants in III-V materials, but implanted Si often shows similar or better activation than implanted group VI dopants such as Se when treated to equilibrium thermal processing[155].

The chemical concentration limit is generally much higher than the electrically active concentration limit of a given dopant for both group IV and group VI species. Electrically active impurity concentration is shown to not exceed $0.5–1.5 \times 10^{19}$ cm^{-3}. This result indicates that both group IV and group VI dopants are electrically limited at high doping levels due to some electrical compensation mechanism. Growth-based dopant incorporation methods have shown much higher (5×10^{19} cm^{-3}) active concentrations[156] but these active concentrations are shown to be metastable in multiple studies[157].

However, Si implantation at room temperature has been shown to be not suitable for III-V fin doping in advanced architectures such as finFET or nanowire FETs due to implant induced damage in narrow III-V fins or wires. Hot implant (I/I–HOT) has been developed and shown to eliminate implant damage in the narrow fins of SOI and bulk Si finFETs[84,85] [see Fig. 23]. This is clearly observed in a series of TEM images shown in Fig. 50 that show that I/I–RT forms an amorphous layer around the fin top/sidewalls[158]. This leads to residual defects after activation anneal [see Fig. 50(b)]. In contrast, I/I-HOT does not form an amorphous layer but maintains excellent crystallinity in both as-implanted (I/I-HOT) and after activation annealed III-V fins [see Fig. 50(c)–(f)]. This is attributed to enhanced annihilation of defects (dynamic annealing) with elevated temperature (i.e. I/I-HOT).

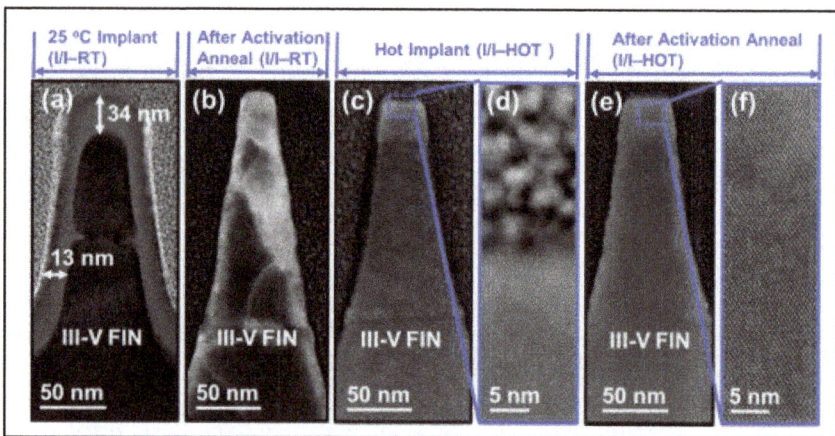

Fig. 50. (a) XTEM image of a III-V fin just after implantation at 25°C (i.e. I/I–RT). An amorphous layer as thick as 34nm is formed after I/I–RT. Fig. 42(b) XTEM image of the I/I–RT fn after activation anneal. Fig. 42(c) and (d) XTEM images of a III-V fin just after hot implant (I/I–HOT). No implant damage is observed and an amorphous layer is not formed after I/I–HOT. Fig. 42(e) and (f) XTEM images of the I/I–HOT III-V fin after activation anneal. Excellent crystallinity is maintained [Source: © 2014, IEEE, Reprinted with permission from Reference 158].

As device dimensions continue to scale, S/D metal contact resistance (i.e. interface resistivity divided by contact area) will continue to increase with the inverse of the S/D contact width. As discussed in previous sections, S/D metal contact interface resistivity is determined by the interface doping concentration. This is limited by dopant solid solubility and the metal barrier height. Since there is an upper limit to the dopant solubility and a lower limit to the achievable contact barrier height, there is a lower limit to the interface resistivity. Furthermore, as the device pitch scales down, so does the contact area, which means that the interface resistivity must scale by at least the same amount in order to preserve the same relative contribution of contact resistance to the total on-state resistance. Eventually, this will no longer be possible due to the limitations mentioned above, at which point the contact resistance is expected to dominate the FET parasitic resistance. Additionally, fin pitch scaling reduces contact area as shown in Fig. 51 and Fig. 52, which will increase contact resistance further. Selective epitaxy in the S/D area could shape the

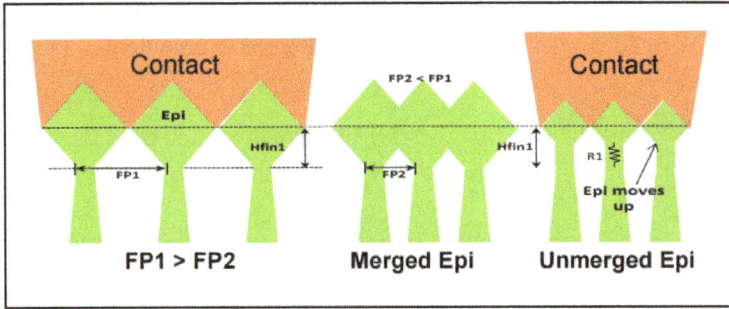

Fig. 51. Fin pitch scaling reduces contact area. Careful design of the epitaxy shape is critical to maximize contact area.

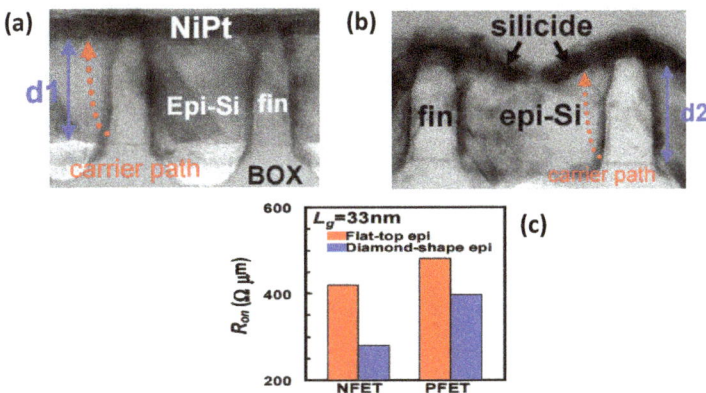

Fig. 52. (a) Merged S/D with flat-top epi. NiPt silicide was formed only on top the S/D surface. (b) Merged fin with diamond-shaped epi. Because of the diamond-shaped epi, the length of current path (d2) gets shorter than d1 in Fig. 51 (a). Also, non-flat surface increases the contact area between Si and silicide. (c) R_{on} comparison between flat-top epi and diamond-shaped epi [Source: © 2009, IEEE, Reprinted with permission from Reference 159].

S/D contact area for subsequent metallization (i.e. metal silicide). For example, epitaxial S/D growth just merging neighboring fin delivers more area for placement of silicide contact than fully merged fins with flat top surface and this will reduce contact resistance[159].

5.7. *Challenges implementing beyond finFET architecture options: Nanowire and vertical transistors*

Although finFETs provide better electrostatic characteristics than planar devices, with continuing gate length scaling beyond the 7nm node with $L_g < 12nm$, short channel control becomes more and more difficult. Especially for bulk finFET, leakage in the region under the fin could dramatically impact the transistor performance.

One of the promising options to combat short channel effects is to use gate-all-around (GAA) devices, such as nanowires and nanosheets, which provide stronger gate control over the channel. Fig. 53 shows the process flow for a nanowire transistor[160]. There are two challenging issues that stand out for nanowire devices:

(1) Inner spacer formation which has to prevent the S/D epi region from being damaged by the SiGe removal process at step 8 as in Fig. 53. Formation of the inner spacer

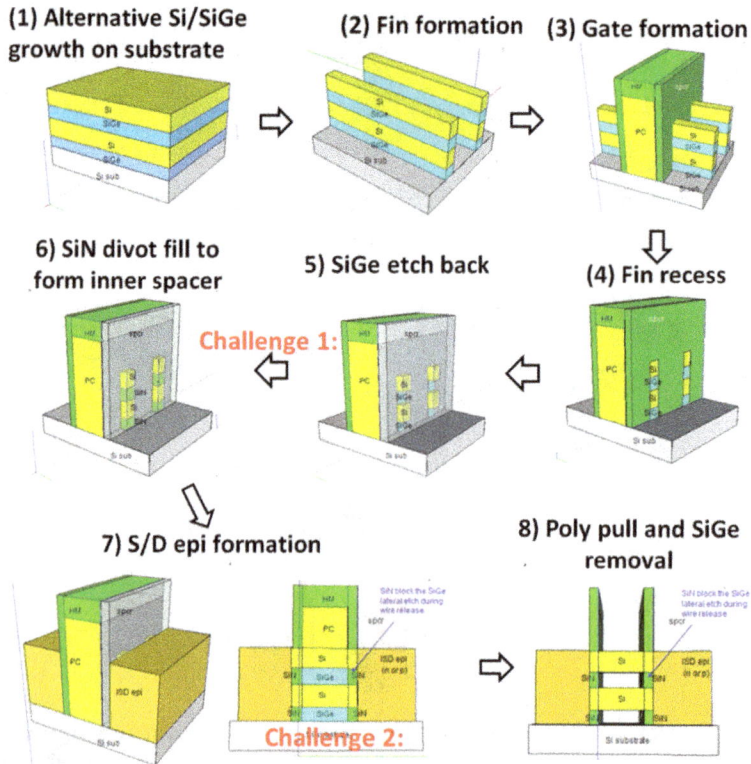

Fig. 53. FEOL process flow for GAA nanowire transistors. Two challenges are highlighted: (1) Inner spacer formation; (2) Parasitic bottom transistors.

involves etching back the SiGe at step 5 and divot filling at step 6. Both these processes (step 5 and step 6) are very hard to control and easily introduce a lot of variations.

(2) Bottom parasitic transistor formation which is naturally formed if the nanowire is built on a bulk substrate. The bottom parasitic transistor should always be in the "off" state. To achieve this, the bottom parasitic transistor region has to be heavily doped. However, heavy doping creates a huge parasitic capacitance which reduces the AC performance. To completely eliminate this issue, either a SOI substrate should be used or some form of dielectric isolation processes needs to be implemented.

Another device option to provide a GAA structure with potentially more aggressive area scaling is the vertical transistor (VFET). Fig. 54 shows the device architecture of VFET[161], it has been shown that ~20% area scaling can be achieved with a VFET-based library rather than a finFET-based library in the 5nm node[162]. The key integration challenges for forming VFET devices are:

(1) Controlling critical dimensions such as gate length and spacer thickness. In conventional horizontal devices, gate length and spacer thickness are usually defined by lithography or ALD processes. Both these process options have very accurate process control and uniformity. However, for VFET, both spacer formation and gate length are largely determined by recess processes that can easily be impacted by etch loading effects.

(2) Achieving different gate lengths in a Vertical FET is difficult and not straight forward, especially the concurrent formation of short channel VFET and long channel VFET is

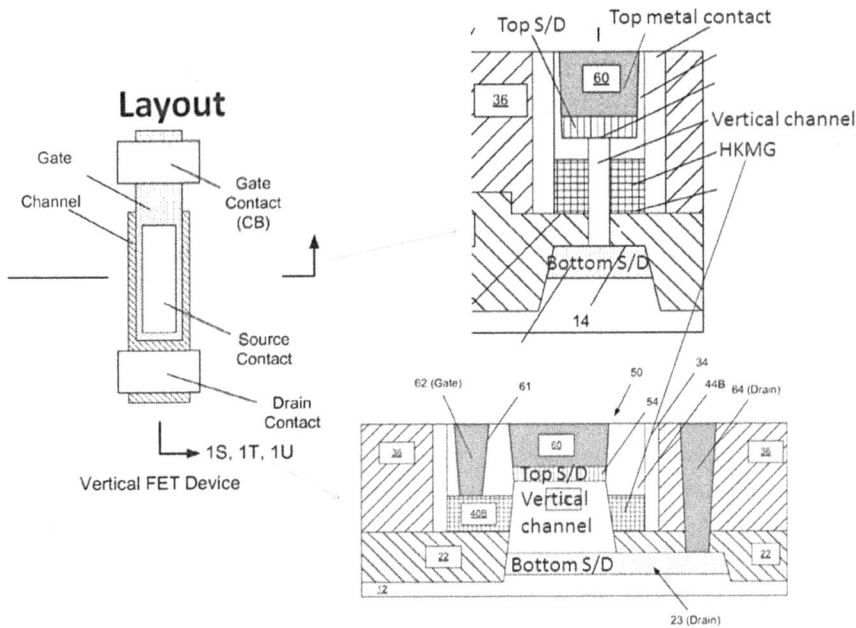

Fig. 54. Vertical transistor device architecture [Source: Reference 161].

difficult. One potential solution is to integrate both VFET and horizontal finFET together, which could provide aggressive area scaling using VFETs while keeping long channel devices on horizontal finFETs[161,163].

6. BEOL Integration Challenges for Interconnect Scaling

The different functions of interconnect mentioned in sections 1 and 3, have different and often conflicting requirements.

Connections between elemental devices (e.g. finFETs or advanced memory elements) tend to be short in length and narrow to achieve packing density, here capacitance (C) is critical. In long range signal routing, resistance (R) and C are important for latency. In the power rails, reliable current carrying capacity is critical. These conflicts often lead to a hierarchical BEOL structure of many metal levels (sometimes over 15!) stacked vertically, and connected by metal vias.

Successful BEOL manufacturing requires control over many contributing factors: overlay of successive layers, critical dimensional non-uniformity, thickness non-uniformity, within-wafer and wafer-to-wafer non-uniformity, as well as composition and adhesion of materials. Advanced node BEOL manufacturing requires several hundred parameters to be within specifications, and scaling continues to tighten these specifications.

In the next several sections, we will show how these challenges are being met, by addressing materials and integration improvements in the components: resistance, capacitance, and finally patterning.

6.1. Challenges in scaling interconnects

Metal line and via resistance (R) depends on the material properties of the conductors and inversely on the dimensions of the metal. Furthermore, scaling tends to increase the electron current density in the metal wires. Collisions of the electrons with metal atoms lead to preferential diffusion of the metal, a phenomenon called electromigration, resulting in voiding and electrical opens. More than a decade ago the industry moved from Aluminum to Copper to contain R and maintain reliability, primarily through reduced electromigration[164,165]. However, at today's very advanced nodes the metal dimensions approach the length of the mean free path of electrons, so scattering effects at line sides and lattice defects cause a non-linear increase in resistance. To improve line resistance the two primary knobs available are: materials and dimensions.

Focusing first on the 'materials' knob, the typical metal line, such as in the 14nm technology, is comprised of multiple layers of material with 2-3 layers surrounding the final electroplated copper center core. These surrounding layers comprise: a barrier which prevents Cu from migrating out into the dielectric or oxygen from migrating into the metal; a liner which facilitates adhesion between the barrier and the Cu; and a Cu seed layer to help initiate the subsequent plating step. To leave as much room as possible for the low-resistivity plated copper fill, the surrounding layers have to be scaled to the minimum thickness required to accomplish their purpose (1-3nm). Still, Fig. 55 shows how, as line

| 64nm Pitch | 45nm Pitch | 34nm Pitch |
| Cu >80% of Area | Cu ~70% | Cu <60% |

Plated Cu

Cu Seed
Liner
Barrier

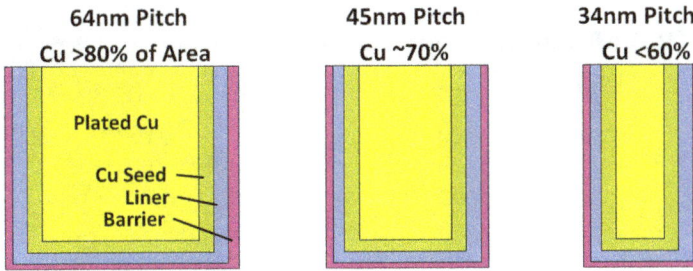

Fig. 55. Low resistance Cu is becoming a smaller portion of the metal line, since the barrier/line/ seed cannot be thinned.

dimensions are shrunk by generation, the plated copper area becomes much less of the total area, causing effective resistance to increase.

Chipmakers are changing these barrier and liner layers[166] either by changing processes from physical vapor deposition (PVD) to atomic layer deposition (ALD) for better thickness control (as already implemented for the TaN barrier), or by changing materials (such as moving the liner from Ta to Co to Ru), or perhaps even by layer elimination (i.e. using a single alloy for the barrier and liner, or eliminating the Cu seed layer and plating directly onto the liner). The latter is enabled by changes in the barrier and liner materials and by moving from electro-plating to electroless-plating. Also there has been an investigation into coatings which can reduce the side-wall scattering of electrons[167]. Clearly, significant changes are underway for fine-pitch metal formation and these changes introduce their own integration concerns such as: chemical mechanical polish without divots, cleans without eroding the new materials, and, perhaps most importantly, reliability characterization concerns such as efficient long-term testing on all of these possible combinations of materials.

Looking now at metal dimension, the only option is to decrease resistance by increasing line height since the line width is constrained by area scaling. Increasing height increases the aspect ratio, which challenges the limits of manufacturing capability: dig a deep, near-90 degree sidewall trench into the interlayer dielectric without damaging the remaining dielectric, fight Van-der-Waals attractions to perfectly clean out the bottom corners, and then do metal fill without pinching off at the top of the line or leaving a void. The etch challenges are being addressed with tool improvements such as 'high frequency pulsed biasing' that, given different charge states on reactants vs products, enables real-time clean-out of etch by-products which, if left unattended, tend to self-limit an etch process. Similarly, wet- and dry-clean processes are continuously being improved by introducing new chemistries, dilutions, and pH realms.

Figure 56 shows how changes in both materials and dimensions can affect the resistance of a thin metal line. The final decision for any given technology is a continuous co-optimization of all of these elements; along with cost and throughput considerations (e.g. ALD has much slower throughput than PVD).

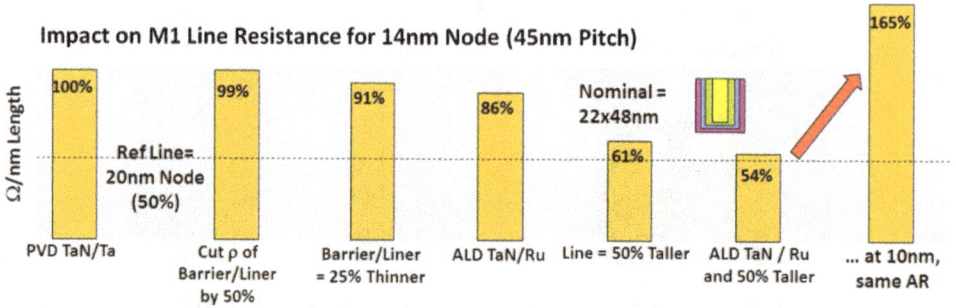

Impact on M1 Line Resistance for 14nm Node (45nm Pitch)

Fig. 56. The resistance of a 22x48nm metal line as a function of material and dimensional changes. Target is to hit the same resistance as the earlier generation (50% of the dumb-shrink to 45nm). Note how the materials changes can make moderate improvements but dimensional changes have a significant impact, Note that even if a solution is found for the current generation, a dumb shrink to next technology node dimensions blows up the problem once again.

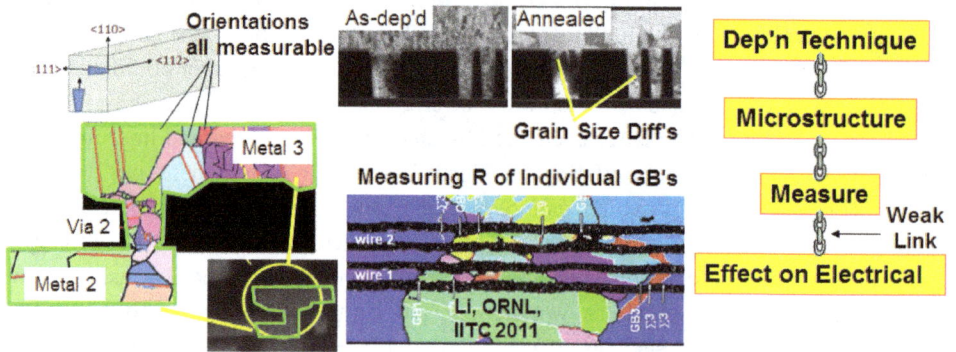

Fig. 57. Drawing the connection from processing to physical microstructure to electrical properties. All components are available, but not yet developed into an efficient feedback/understanding.

Looking at even finer detail of resistance, we recognize that resistance happens at the atomic and grain-size level, due to disruptions in bonding at grain boundaries. Analysis of the distribution of grain and grain boundary orientations is possible, so the impact of processing (e.g. depositions and anneals) on this microstructure can be determined. Finally, with careful measurements, the connection between microstructure and electrical behavior can be studied. This completes the chain from dimensions to deposition and processing to microstructure and finally electrical behavior, with the last link representing our weakest understanding. See Fig. 57.

This understanding of the importance of microstructure is also leading to possible changes in the middle of line (MOL) where the Cu/Ox world of the BEOL must meet up with the Si/HK/MG world of the FEOL. The historical favorite, W, is being challenged by other pure metals such as Cobalt and Ruthenium, which, though in bulk form have higher resistivities, at these small dimensions have lower resistivities due to their intrinsically different grain sizes and grain boundaries. The MOL is rapidly playing a much larger role in any integration, especially as we look to newer device architectures such as VFETs. The ability to extract the current from the device and deliver it to the BEOL with a minimum of resistance loss is a fundamental factor in selecting next generation technologies.

6.2. *Challenges in reducing BEOL capacitance*

Soon after moving to Cu, the industry also moved from SiO_x to low-k and ultra-low-k (ULK) dielectrics in an effort to contain C increase. However, such materials are more susceptible to early dielectric wear-out, called Time Dependent Dielectric Breakdown TDDB, in which hot electrons are injected into the dielectric by the E-field between metal features. Unfortunately, scaling increases the strength of the E-field, and accelerates TDDB. To further reduce C, ultra-low-k materials have been developed, which include open space, in form of pores, within the material, but this decreases their mechanical strength (see Fig. 58).

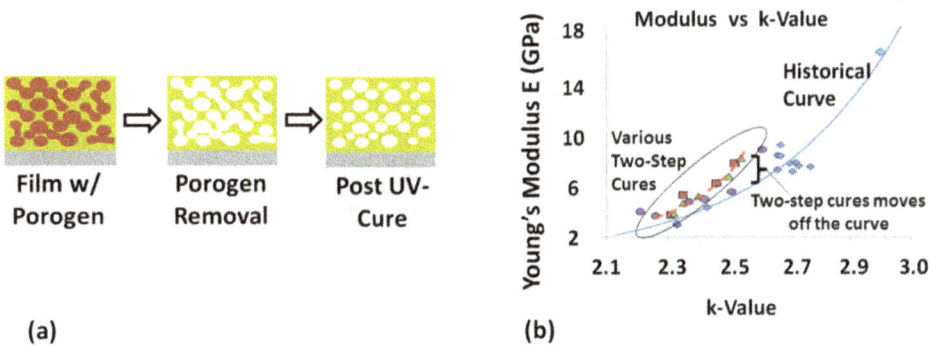

Fig. 58. (a) Typical processing of an ultra-low-k dielectric, (b) Young's Modulus vs k-value showing the impact of process induced variations on the mechanical strength (important for enduring the stresses of full packaging of a chip). Multi-step cures enable improvements in final properties (stronger/stiffer for a given k-value).

The introduction of new materials to reduce capacitance has almost stopped, due to these issues of mechanical stability and strength. In response, there has been considerable effort to increase the mechanical strength of porous materials by atomic bond engineering within the dielectrics[168].

Once the blanket dielectric is formed, it must be etched to allow the damascene processing (i.e. the metal fill and polish) to occur. This etching can be quite damaging to the dielectric, effectively negating some of the efforts to get to lower k (see Fig. 59). For this reason, one approach for obtaining lower BEOL capacitance of selected metal levels is to completely remove the dielectric between parallel lines and then to seal it across the top with the next level of dielectric, leaving a void, or "air gap" between the lines.

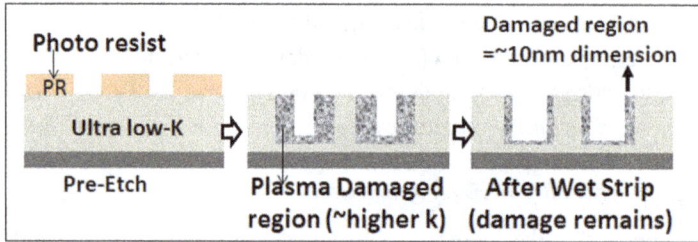

Fig. 59. Typical reactive ion etching of trenches into dielectric (where the metal lines will eventually fill) results in several nm of damaged dielectric (with higher k) remaining. This means, in next generation devices, as the dimensions scales, plasma damaged region will be a larger % of the final dielectric thickness.

6.3. *Challenges implementing design and patterning*

Independent of material and structural enhancements, there are integration enhancements possible, but these often require a coupling back to the design world. As described in section 4, for developing nodes there is a shift toward managing designs through more restrictive ground rules. This is necessary partly because lithography has been forced to remain at 193nm wavelength for several nodes, and desired CD's are well below the resolution limits of these 193nm lithography tools. After extending the lithography tool's resolving power through increasingly complex and restrictive RET to their absolute resolution limit in the 22nm technology node, further pitch scaling was only possible by means of multiple exposure patterning. Two different multiple exposure patterning approaches have been developed and are heavily used in advanced technology nodes today. One employs multiple sequential lithography steps that are optically decoupled by memorizing each exposure in a sacrificial film stack and then transferring the collective shapes into the dielectric. The repetitious sequence of lithography and etch steps gave this technique its name: Litho-Etch-Litho-Etch (LELE) (see Fig. 60). LELE has the drawback of requiring multiple expensive passes through the lithography tools, and of lowering yield due to tight overlay requirements at each metal level. The other multiple exposure

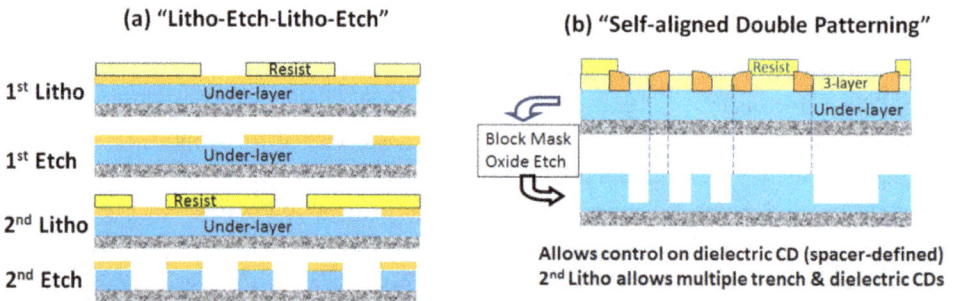

Fig. 60. Two different approaches (a) Litho-Etch-Litho-Etch (LELE) and (b) Self-aligned double patterning (SADP) for obtaining lines and spaces in the BEOL which are below that obtainable from a single lithography step.

patterning solution is borrowed from FEOL patterning techniques: Self-aligned double patterning (SADP)[169,170]. This method uses spacers deposited on the sides of a wafer relief feature to effectively double the pitch of the etched features that become metal lines. However, SADP does not cope well with interconnect layouts that turn corners, so designs have moved to wiring that is almost entirely unidirectional in each metal level and orthogonal in successive levels.

6.4. *Packaging*

Due to the many conflicting needs described above, scaling the BEOL is a challenge of Thermo-Mechanical-Electrical co-optimization. There is considerable effort to reduce the stress on the BEOL due to packaging by extensive mechanical modelling, managing materials selection within package, managing the transitions between successive layers in the interconnect stack, and by inserting drawn features on the chip to deflect or arrest cracks. In fact the co-optimization extends to the FEOL devices also. Part of the reason TDDB becomes more of an issue with scaling, is that operating voltages do not scale as fast as dimensions, which intensifies E-fields within the chip. Therefore, adopting devices which can operate at lower V_{dd} would ease the reliability burden of BEOL dielectric wear-out.

At the same time, the BEOL is part of the packaged chip and the use of plastic packages results in significant stress on the BEOL, between the Si and the package, resulting in cracking and failure. This stress is due to the very different coefficients of thermal expansion of the package compared to Si. Moreover, the development of mobile applications has driven thinner package form-factors requiring thinner die, which can lead to increased local strain of the BEOL, and early failure.

7. Conclusion

This review has discussed device and integration challenges that prompted innovative solutions to continue the power, performance and area scaling of CMOS integrated chips down to the non-planar device regime. Device and integration challenges have made planar bulk transistor technology uncompetitive beyond the 22nm technology node. Innovative solutions have prompted the industry to adapt non-planar finFET architecture beyond the 22nm technology node. With non-planar architectures, issues related to patterning and integration supersedes device related fundamental challenges and design technology co-optimization becomes necessary. This becomes more important as the technology scales further down below sub 10nm node. So far the industry has adapted various techniques such as SADP and SAQP to surpass the challenges in patterning. However, fin pitch, gate pitch and metal pitch reduction along with filing the nanometric gaps with either metals or dielectrics continue to remain as a deterrent in adapting aggressive technology nodes. This also means that as the pitches reduce, selective deposition of films in the monolayer regime and selective etching of films to the nanometer precision are very important. This review discussed in depth challenges associated in introducing non-planar technology and in particular finFET formation, spacer and dummy gate formation, source drain epi formation, SAC and RMG, diffusion break, contact, interconnect design and RC delay, patterning, and packaging. It was shown that that several solutions that helped scaling planar technology do not give an equal performance improvement in non-planar technology and innovative solutions are required to maintain the PPA trade-off and manufacturing yield. In addition, this review also highlighted the challenges in introducing alternate channel materials into non-planar devices and alternate device architecture beyond finFETs.

References

1. International Technology Roadmap for Semiconductors 2.0 (2015).
2. S. Borkar, Design Challenges of Technology Scaling, *IEEE Micro* **19**(4), 23-29 (1999).
3. C. Z. Sze and Kwok K. Ng, *Physics of Semiconductor Devices*, 3rd Edition (Wiley 2006).
4. K. Roy, S. Mukhopadhyay and H. Mahmoodi-Meimand, Leakage current mechanisms and leakage reduction techniques in deep-submicrometer CMOS circuits, *Proceedings of the IEEE*, **91**(2), 305-327 (2003).
5. A. Khakifirooz, K. Cheng, T. Nagumo, N. Loubet, T. Adam, A. Reznicek, J. Kuss, D. Shahrjerdi, R. Sreenivasan, S. Ponoth, H. He, P. Kulkarni, Q. Liu, P. Hashemi, P. Khare, S. Luning, S. Mehta, J. Gimbert, Y. Zhu, Z. Zhu, J. Li, A. Madan, T. Levin, F. Monsieur, T. Yamamoto, S. Naczas, S. Schmitz, S. Holmes, C. Aulnette, N. Daval, W. Schwarzenbach, B. Y. Nguyen, V. Paruchuri, M. Khare, G. Shahidi and B. Doris, Strain engineered extremely thin SOI (ETSOI) for high-performance CMOS, *Proceedings of 2012 Symposium on VLSI Technology (VLSIT)*, 117-118 (2012).
6. Y.-K. Choi, K. Asano, N. Lindert, V. Subramanian, T.-J. King, J. Bokor and C. Hu, Ultrathin-body SOI MOSFET for deep-sub-tenth micron era, *IEEE Elec. Dev. Lett.*, **21**(5), 254-255 (2000).
7. Y.-K. Choi, D. Ha, T.-J. King and C. Hu, Nanoscale ultrathin body PMOSFETs with raised selective Germanium source/drain, *IEEE Elec. Dev. Lett.*, **22**(9), 447-448 (2001).
8. S. A. Vitale, P. W. Wyatt, N. Checka, J. Kedzierski, and C. L. Keast, FDSOI process technology for subthreshold operation ultralow-power electronics, *Proceedings of the IEEE*, **98**(2), 333-342 (2010).
9. D. J. Schepis, F. Assaderaghi, D. S. Yee, W. Rausch, R. J. Bolam, A. C. Ajmera, E. Leobandung, S. B. Kulkarni, R. Flaker, D. Sadana, H. J. Hovel, T. Kebede, C. Schiller, S. Wu, L. F. Wagner, M. J. Saccamango, S. Ratanaphanyarat, J. B. Kuang, M. C. Hsieh, K. A. Tallman, R. M. Martino, D. Fitzpatrick, D. A. Badami, M. Hakey, S. F. Chu, B. Davari, and G. G. Shahidi, A 0.25 μm CMOS on SOI and its application to 4 Mb SRAM, IEEE International Electron Devices Meeting, *IEDM. Tech. Dig.*, 587-590 (1997).
10. M. R. Casu, G. Masera, C. Piccinini, M. R. Roch and M. Zamboni, Comparative analysis of PD-SO1 active body-biasing circuits, *IEEE International SO1 Conference*, 94-95 (2000).
11. C.-T. Chuang, P.-Fe Lu and C. J. Anderson, SOI for digital CMOS VLSI : design considerations and advances, *Proceedings of the IEEE*, **86**(4) 689-720 (1998).
12. K. Cheng, S. Seo, J. Faltermeier, D. Lu, T. Standaert, I. Ok, A. Khakifirooz, R. Vega, T. Levin, J. Li, J. Demarest, C. Surisetty, D. Song, H. Utomo, R. Chao, H. He, A. Madan, P. DeHaven, N. Klymko, Z. Zhu, S. Naczas, Y. Yin, J. Kuss, A. Jacob, D. Bae, K. Seo, W. Kleemeier, R. Sampson, T. Hook, B. Haran, G. Gifford, D. Gupta, H. Shang, H. Bu, M. Na, P. Oldiges, T. Wu, B. Doris, K. Rim, E. Nowak, R. Divakaruni and M. Khare, IEEE Bottom oxidation through STI (BOTS) - A novel approach to fabricate dielectric isolated FinFETs on bulk substrates, *Symposium on VLSI Technology (VLSI-Technology): Digest of Technical Papers*, 1-2 (2014).

13. R. H. Yan, A. Ourmazd and K. F. Lee, Scaling the Si MOSFET: from bulk to SOI to bulk, *IEEE Trans. on Elec. Dev.*, **39**(7), 1704-1710 (1992).
14. L. Geppert, The amazing vanishing transistor act, *IEEE Spectrum*, **39**(10), 28-33 (2002).
15. X. Huang, W.-C. Lee, C. Kuo, D. Hisamoto, L. Chang, J. Kedzierski, E. Anderson, H. Takeuchi, Y.-K. Choi, K. Asano, V. Subramanian, T.-J. King, J. Bokor, and C. Hu, Sub 50-nm FinFET: PMOS, *IEEE International Elec. Dev. Meeting (IEDM) Tech. Dig.*, 67-70 (1999).
16. Y.-K. Choi, N. Lindert, P. Xuan, S. Tang, D. Ha, E. Anderson, T.-J. King, J. Bokor and C. Hu, Sub-20nm CMOS FinFET technologies, *IEEE International Elec. Dev. Meeting Tech. Dig.*, 421-424 (2001).
17. B. Yu, L. Chang, S. Ahmed, H. Wang, S. Bell, C.-Y. Yang, C. Tabery, C. Hu, T.-J. King, J. Bokor, M.-R. Lin, and D. Kyser, FinFET scaling to 10nm gate length, *IEEE International Elec. Dev. Meeting Tech. Dig.*, 251-254 (2002).
18. B. S. Doyle, S. Datta, M. Doczy, S. Hareland, B. Jin, J. Kavalieros, T. Linton, A. Murthy, R. Rios and R. Chau, High Performance Fully-Depleted Tri-Gate CMOS Transistors, *IEEE Elec. Dev. Lett.*, **24**(4), 263-265 (2003).
19. S. Migita, Y. Morita, T. Matsukawa, M. Masahara and H. Ota, Experimental Demonstration of Ultrashort-Channel (3 nm) Junctionless FETs Utilizing Atomically Sharp V-Grooves on SOI, *IEEE Transactions on Nanotechnology*, **13**, 208-215 (2014).
20. S. B. Desai, S. R. Madhvapathy, A. B. Sachid, J. P. Llinas, Q. Wang, G. H. Ahn, G. Pitner, M. J. Kim, J. Bokor, C. Hu, H.-S. P. Wong and A. Javey, MoS2 transistors with 1-nanometer gate lengths, *Science*, **354**(6308), 99-102 (2016).
21. A. P. Jacob, Investigation of Future Nanoscaled Semiconductor Heterostructures and CMOS Devices, PhD Thesis, Chalmers University of Technology and Gothenburg University, Sweden, ISBN 91-628-5464-X (2002).
22. A. P. Jacob, T. Myrberg, O. Nur, M. Willander, P. Lundgren, E. Ö. Sveinbjörnsson, L. L. Ye, A. Thölen and M. Caymax, Cryogenic performance of ultrathin oxide MOS capacitors with in situ doped p+ poly-Si$_{1-x}$Ge$_x$ and poly-Si gate materials, *Semicond. Sci. and Tech.*, **17**(9), 942-946 (2002).
23. K. Mistry, C. Allen, C. Auth, B. Beattie, D. Bergstrom, M. Bost, M. Brazier, M. Buehler, A. Cappellani, R. Chau, C.-H. Choi, G. Ding, K. Fischer, T. Ghani, R. Grover, W. Han, D. Hanken, M. Hattendorf, J. He, J. Hicks, R. Huessner, D. Ingerly, P. Jain, R. James, L. Jong, S. Joshi, C. Kenyon, K. Kuhn, K. Lee, H. Liu, J. Maiz, B. McIntyre, P. Moon, J. Neirynck, S. Pae, C. Parker, D. Parsons, C. Prasad, L. Pipes, M. Prince, P. Ranade, T. Reynolds, J. Sandford, L. Shifren, J. Sebastian, J. Seiple, D. Simon, S. Sivakumar, P. Smith, C. Thomas, T. Troeger, P. Vandervoorn, S. Williams and K. Zawadzki, A 45nm Logic Technology with High-k+Metal Gate Transistors, Strained Silicon, 9 Cu Interconnect Layers, 193nm Dry Patterning, and 100% Pb-free Packaging, *IEEE International Elec. Dev. Meet.*, 247-250 (2007).
24. K. Henson, H. Bu, M. H. Na, Y. Liang, U. Kwon, S. Krishnan, J. Schaeffer, R. Jha, N. Moumen, R. Carter, C. DeWan, R. Donaton, D. Guo, M. Hargrove, W. He, R. Mo, R. Ramachandran, K. Ramani, K. Schonenberg, Y. Tsang, X. Wang, M. Gribelyuk, W. Yan, J. Shepard, E. Cartier, M. Frank, E. Harley, R. Arndt, R. Knarr, T. Bailey, B. Zhang, K. Wong, T. Graves-Abe, E. Luckowski, D.-G. Park, V. Narayanan, M. Chudzik, and M. Khare, Gate Length Scaling and High Drive Currents Enabled for High Performance SOI Technology using High-k/Metal Gate, *IEEE International Elec. Dev. Meeting (IEDM) Tech. Dig.*, 645-648 (2008).
25. L.-Å. Ragnarsson, Z. Li, J. Tseng, T. Schram, E. Rohr, M. J. Cho, T. Kauerauf, T. Conard, Y. Okuno, B. Parvais, P. Absil, S. Biesemans, and T. Y. Hoffmann, Ultralow-EOT (5 Å) Gate-First and Gate-Last High Performance CMOS Achieved by Gate-Electrode Optimization, *IEEE International Elec. Dev. Meet.(IEDM) Tech. Dig.*, 663-666 (2009).

26. T. Ghani, M. Armstrong, C. Auth, M. Bost, P. Charvat, G. Glass, T. Hoffiann', K. Johnson', C. Kenyon, J. Klaus, B. Mclntyre, K. Mistry, A. Murthy, 1. Sandford, M. Silberstein, S. Sivakumar, P. Smith, K. Zawadzki, S. Thompson and M. Bohr, A 90nm high volume manufacturing logic technology featuring novel 45nm gate length strained silicon CMOS transistors, *IEEE International Elec. Dev. Meet.(IEDM) Tech. Dig.*, 11.6.1-11.6.3 (2003).

27. E. Y. Wu, R. P. Vollertsen, R. Jarnmy, A. Strong and C. Radens, Leakage current and reliability evaluation of ultra-thin reoxidized nitride and comparison with silicon dioxides, *40th Annual Reliability Phys. Symp. Proc.*, 255-267 (2002).

28. P. Bai, C. Auth, S. Balakrishnan, M. Bost, R. Brain, V. Chikarmane, R. Heussner, M. Hussein, J. Hwang, D. Ingerly, R. James, J. Jeong, C. Kenyon, E. Lee, S.-H. Lee, N. Lindert, M. Liu, Z. Ma, T. Marieb, A. Murthy, R. Nagisetty, S. Natarajan, J. Neirynck, A. Ott, C. Parker, J. Sebastian, R. Shaheed, S. Sivakumar, J. Steigerwald, S. Tyagi, C. Weber, B. Woolery, A. Yeoh, K. Zhang, and M. Bohr, A 65nm Logic Technology Featuring 35nm Gate Lengths, Enhanced Channel Strain, 8 Cu Interconnect Layers, Low-k ILD and 0.57 μm2 SRAM Cell, *IEEE International Elec. Dev. Meet.(IEDM) Tech. Dig.*, 657-660 (2004).

29. C. Prasad, M. Agostinelli, C. Auth, M. Brazier, R. Chau, G. Dewey, T. Ghani, M. Hattendorf, J. Hicks, J. Jopling, J. Kavalieros, R. Kotlyar, M. Kuhn, K. Kuhn, J. Maiz, B. McIntyre, M. Metz, K. Mistry, S. Pae, W. Rachmady, S. Ramey, A. Roskowski, J. Sandford, C. Thomas, C. Wiegand, and J. Wiedemer, Dielectric Breakdown in a 45 nm High-K/Metal Gate Process Technology, *IEEE CFP08RPS-CDR 46th Annual International Reliability Phys. Symp.*, Phoenix, 667-668 (2008).

30. S. Salahuddin and S. Datta, Use of negative capacitance to provide voltage amplification for low power nanoscale devices, *Nano. Lett.*, **8**(2), 405-410 (2008).

31. C. W. Yeung, A. I. Khan, J.-Y. Cheng, S. Salahuddin and C. Hu, Non-Hysteretic Negative Capacitance FET with Sub- 30mV/dec Swing over 106X Current Range and ION of 0.3mA/μm without Strain Enhancement at 0.3V VDD, *The International Conference on Simulations of Semiconductor Processes and Devices (SISPAD)*, 257-259 (2012).

32. C. H. Cheng and A. Chin, Low-Voltage Steep Turn-On pMOSFET Using Ferroelectric Highk Gate Dielectric, *IEEE Elec. Dev. Lett.*, **35**(2), 274-276 (2014).

33. G. A. Salvatore, D. Bouvet and A. M. Ionescu, Demonstration of subthrehold swing smaller than 60mV/decade in Fe-FET with P (VDF-TrFE)/SiO 2 gate stack, *IEEE International Elec. Dev. Meet. (IEDM) Tech. Dig.*, 1-4 (2008).

34. K.-S. Li, P.-G. Chen, T.-Y. Lai, C.-H. Lin, C.-C. Cheng, C.-C. Chen, Y.-J. Wei, Y.-F. Hou, M.-H. Liao, M.-H. Lee, M.-C. Chen, J.-M. Sheih, W.-K. Yeh, F.-L. Yang, S. Salahuddin and C. Hu, Sub-60mV-swing negative-capacitance FinFET without hysteresis, *IEEE International Elec. Dev. Meet. (IEDM) Tech. Dig.*, 22.6.1-22.6.4 (2015).

35. P. Polakowski and J. Müller, Ferroelectricity in undoped hafnium oxide, *App. Phys. Lett.*, **106**(23), 232905 (2015).

36. J. Muller, T. S. Boscke, S. Muller, E. Yurchuk, P. Polakowski, J. Paul and D. Martin, Ferroelectric hafnium oxide: A CMOS-compatible and highly scalable approach to future ferroelectric memories, *IEEE International Elec. Dev. Meet. (IEDM) Tech. Dig.*, 10.8.1-10.8. 4 (2013).

37. J. Müller, T. S. Böscke, U. Schröder, S. Mueller, D. Bräuhaus, U. Böttger, L. Frey and T. Mikolajick, Ferroelectricity in simple binary ZrO_2 and HfO_2, *Nano letters*, **12**(8), 4318-4323 (2012).

38. A. I. Khan, K. Chatterjee, B. Wang, S. Drapcho, L. You, C. Serrao, S. R. Bakaul, R. Ramesh and S. Salahuddin, Negative capacitance in a ferroelectric capacitor, *Nature Materials*, **14**, 182-186 (2015).

39. R. Materlik, C. Künneth and A. Kersch, The origin of ferroelectricity in $Hf_{1-x}Zr_xO_2$: A computational investigation and a surface energy model, *J. App. Phys.*, **117**, 134109.1-134109.15 (2015).
40. F. A. McGuire, Z. Cheng, K. Price and A. D. Franklin, Sub-60 mV/decade switching in 2D negative capacitance field-effect transistors with integrated ferroelectric polymer, *App. Phys. Lett.*, **109**, 093101.1-093101.5 (2016).
41. E. Yurchuk, J. Müller, S. Knebel, J. Sundqvist, A. P. Graham and T. Melde, Impact of layer thickness on the ferroelectric behaviour of silicon doped hafnium oxide thin films, *Thin Solid Films*, **533**, 88-92 (2013).
42. G. Sun, Y. Sun, T. Nishida, and S. E. Thompson, Hole mobility in silicon inversion layers: Stress and surface orientation, *J. App. Phys.*, **102**, 084501.1-084501.7 (2007).
43. S. E. Thompson, M. Armstrong, C. Auth, M. Alavi, M. Buehler, R. Chau, S. Cea, T. Ghani, G. Glass, T. Hoffman, C.-H. Jan, C. Kenyon, J. Klaus, K. Kuhn, Zhiyong Ma, B. Mcintyre, K. Mistry, A. Murthy, B. Obradovic, R. Nagisetty, Phi Nguyen, S. Sivakumar, R. Shaheed, L. Shifren, B. Tufts, S. Tyagi, M. Bohr and Y. El-Mansy, A 90-nm Logic Technology Featuring Strained-Silicon, *IEEE Trans. Elec. Dev.*, **51**(11), 1790-1797 (2004).
44. S. Ito, H. Namba, K. Yamaguchi, T. Hirata, K. Ando, S. Koyama, S. Kuroki, N. Ikezawa, T. Suzuki, T. Saitoh and T. Horiuchi, Mechanical stress effect of etch-stop nitride and its impact on deep submicrometer transistor design, in *IEEE Elec. Dev. Meet. (IEDM) Tech. Dig.*, 247-250 (2000).
45. A. Shimizu, K. Hachimine, N. Ohki, H. Ohta, M. Koguchi, Y. Nonaka, H. Sato and F. Ootsuka, A Local mechanical-stress control (LMC): a new technique for CMOS-performance enhancement, *IEEE Elec. Dev. Meet. (IEDM) Tech. Dig.*, 433-436 (2001).
46. C.-H. Chen, T. L. Lee, T. H. Hou, C. L. Chen, C. C. Chen, J. W. Hsu, K. L. Cheng, Y. H. Chiu, H. J. Tao, Y. Jin, C. H. Diaz, S. C. Chen, and M.-S. Liang, Stress Memorization Technology (SMT) by Selectively Strained-Nitride Capping for Sub-65nm High-Performance Strained-Si Device Application, *IEEE Symposium on VLSI Tech, Digest of Tech. Papers*, 56-57 (2004).
47. K.-W. Ang, K.-J. Chui, V. Bliznetsov, Y. Wang, L.-Y. Wong, C.-H Tung, N. Balasubramanian, M.-Fu Li, G. Samudra and Y.-C. Yeo, Thin body silicon-on-insulator N-MOSFET with silicon-carbon source/drain regions for performance enhancement, *IEEE International Elec. Dev. Meet. (IEDM) Tech. Dig.*, 497-500 (2005).
48. E. Parton and P. Verheyen, Strained silicon — the key to sub-45 nm CMOS, *III-Vs Review*, **19**(3), 28-31 (2006).
49. M. Cai, K. Ramani, M. Belyansky, B. Greene, D. H. Lee, S. Waidmann, F. Tamweber and W. Henson, Stress liner effects for 32-nm SOI MOSFETs with HKMG, *IEEE Trans. Elec. Dev.*, **57**(7), 1706-1709 (2010).
50. T. Satô, Y. Takeishi and H. Hara, Effects of Crystallographic Orientation on Mobility, Surface State Density, and Noise in p-Type Inversion Layers on Oxidized Silicon Surfaces, *Jap. J. App. Phys.*, **8**(5), 1347-4065 (1969).
51. B. Mereu, C. Rossel, E. P. Gusev and M. Yang, The role of Si orientation and temperature on the carrier mobility in metal oxide semiconductor field effect transistors with ultrathin HfO_2 gate dielectrics, *J. App. Phys.*, **100**, 014504.1-014504.6 (2006).
52. M. E. Levinshtein and S. L. Rumyantsev, *Handbook Series on Semiconductor Parameters*, **1**, 1-32 (World Scientific, London, 1996).
53. M. P. Mikhailova *Handbook Series on Semiconductor Parameters*, **1**, 147-168 (World Scientific, London, 1996).
54. L. E. Vorobyev, *Handbook Series on Semiconductor Parameters*, **1**, 33-57 (World Scientific, London, 1996).

55. Y. A. Goldbery, *Handbook series on semiconductor parameters*, **1**, 191-213 (World Scientific, London, 1996).
56. R. J. W Hill, C. Park, J. Barnett, J. Price, J. Huang, N. Goel, W. Y. Loh, J. Oh, C. E. Smith, P. Kirsch, P. Majhi and R. Jammy, Self-aligned III-V MOSFETs heterointegrated on a 200 mm Si substrate using an industry standard process flow, *IEEE International Elec. Dev. Meet. (IEDM) Tech. Dig.*, 6.2.1-6.2.4. (2010).
57. J.-H. Hur and S. Jeon, III–V compound semiconductors for mass-produced nano-electronics: theoretical studies on mobility degradation by dislocation, *Scientific Reports*, **6**, 22001 (2016).
58. T Myrberg, AP Jacob, O Nur, M Friesel, M Willander, CJ Patel, Y Campidelli, C Hernandez, O Kermarrec and D Bensahel, Structural properties of relaxed Ge buffer layers on Si (0 0 1): effect of layer thickness and low temperature Si initial buffer, *J. Mat. Sci: Mat. in Elec.*, **15**(7), 411-417 (2004).
59. S. Datta, J. Brask, G. Dewey, M. Doczy, B. Doyle, B. Jin, J. Kavalieros, M. Metz, A. Majumdar, M. Radosavljevic and R. Chau, Advanced Si and SiGe Strained Channel NMOS and PMOS Transistors with High-K/Metal-Gate Stack, *Proceedings of the IEEE - Bipolar/BiCMOS Circuits and Technology meeting*, 194-197 (2004).
60. S. Krishnan, U. Kwon, N. Moumen, M. W. Stoker, E. C. T. Harley, S. Bedell, D. Nair, B. Greene, W. Henson, M. Chowdhury, D. P. Prakash, E. Wu, D. Ioannou, E. Cartier, M.-H. Na, S. Inumiya, K. Mcstay, L. Edge, R. Iijima, J. Cai, M. Frank, M. Hargrove, D. Guo, A. Kerber, H. Jagannathan, T. Ando, J. Shepard, S. Siddiqui, M. Dai, H. Bu, J. Schaeffer, D. Jaeger, K. Barla, T. Wallner, S. Uchimura, Y. Lee, G. Karve, S. Zafar, D. Schepis, Y. Wang, R. Donaton, S. Saroop, P. Montanini, Y. Liang, J. Stathis, R. Carter, R. Pal, V. Paruchuri, H. Yamasaki, J.-H. Lee, M. Ostermayr, J.-P. Han, Y. Hu, M. Gribelyuk, D.-G. Park, X. Chen, S. Samavedam, S. Narasimha, P. Agnello, M. Khare, R. Divakaruni, V. Narayanan and M. Chudzik, A manufacturable dual channel (Si and SiGe) high-k metal gate CMOS technology with multiple oxides for high performance and low power applications, *IEEE International Elec. Dev. Meet. (IEDM) Tech. Dig.*, 28.1.1-28.1.4 (2011).
61. C. Le Royer, A. Villalon, M. Cassé, D. Cooper, J. Mazurier, B. Prévitali, C. Tabone, P. Perreau, J.-M. Hartmann, P. Scheiblin, F. Allain, F. Andrieu, O. Weber, P. Batude, O. Faynot and T. Poiroux, First demonstration of ultrathin body c-SiGe channel FDSOI pMOSFETs combined with SiGe(:B) RSD: Drastic improvement of electrostatics (Vth,p tuning, DIBL) and transport (µ0, Isat) properties down to 23nm gate length, *IEEE International Elec. Dev. Meet. (IEDM) Tech. Dig.*, 16.5.1-16.5.4 (2011).
62. K. Cheng, A. Khakifirooz, N. Loubet, S. Luning, T. Nagumo, M. Vinet, Q. Liu, A. Reznicek, T. Adam, S. Naczas, P. Hashemi, J. Kuss, J. Li, H. He, L. Edge, J. Gimbert, P. Khare, Y. Zhu, Z. Zhu, A. Madan, N. Klymko, S. Holmes, T. M. Levin, A. Hubbard, R. Johnson, M. Terrizzi, S. Teehan, A. Upham, G. Pfeiffer, T. Wu, A. Inada, F. Allibert, B.-Y. Nguyen, L. Grenouillet, Y. Le Tiec, R. Wacquez, W. Kleemeier, R. Sampson, R. H. Dennard, T. H. Ning, M. Khare, G. Shahidi and B. Doris, High performance extremely thin SOI (ETSOI) hybrid CMOS with Si channel NFET and strained SiGe channel PFET, *IEEE International Elec. Dev. Meet. (IEDM) Tech. Dig.*, 18.1.1-18.1.4 (2012).
63. M.-H. Chiang, J.-N. Lin, K. Kim, and C.-T. Chuang, Random Dopant Fluctuation in Limited-Width FinFET Technologies, *IEEE Trans. Elec. Dev.*, **54**(8), 2055-2060 (2007).
64. H.-J. Li, P. Kohli, S. Ganguly, T. A. Kirichenko, P. Zeitzoff, K. Torres and S. Banerjee, Boron diffusion and activation in the presence of other species, *IEEE International Elec. Dev. Meet. (IEDM) Tech. Dig.*, 515-518 (2000).
65. Y. Nishi, Y. Tsuchiya, A. Kinoshita, T. Yamauchi, and J. Koga, Interfacial Segregation of Metal at NiSi/Si Junction for Novel Dual Silicide Technology, *IEEE International Elec. Dev. Meet. (IEDM) Tech. Dig.*, 135-138 (2007).

66. J. D. Yearsley, J. C. Lin, E. Hwang, S. Datta, and S. E. Mohney, Ultra low-resistance palladium silicide Ohmic contacts to lightly doped n-InGaAs, *J. Appl. Phys.* **112**, 054510 (2012).
67. E. Huang, E. Joseph, H. Bu, X. Wang, N. Fuller, C. Ouyang, E. Simonyi, H. Shobha, T. Cheng, A. Mallikarjunan, I. Lauer, S. Fang, W. Haensch, C.-Y Sung, S. Purushothaman, and G. Shahidi, Low-k Spacers for Advanced Low Power CMOS Devices with Reduced Parasitic Capacitances, *IEEE Inter. SOI Conf. Proc.*, 19-20 (2008)
68. T. Yamashita, S. Mehta, V.S. Basker, R. Southwick, A. Kumara, R. Kambhampatib, R. Sathiyanarayanana, J. Johnsona, T. Hook, S. Cohen, J. Li, A. Madan, Z. Zhu, L. Tai, Y. Yao, P. Chinthamanipeta, M. Hopstaken, Z. Liu, D. Lu, F. Chena, S. Khana, D. Canaperi, B. Haran, J. Stathis, P. Oldiges, C.-H. Lin, S. Narasimhaa, A. Bryant, W. K. Hensona, S. Kanakasabapathy, K.V.R.M. Muralia, T. Gow, D. McHerron, H. Bu and M. Khare, A Novel ALD SiBCN Low-k Spacer for Parasitic Capacitance Reduction in FinFETs, *Symp. on VLSI Tech.y – Dig. Tech. Papers*, T154-T155 (2015).
69. P. J. M. Havinga, Mobile Multimedia Systems, Ph.D. Thesis University of Twente, ISBN 90-365-1406-1 (2000).
70. P. Kogge, K. Bergman, S. Borkar, D. Campbell, W. Carlson, W. Dally, M. Denneau, P. Franzon, W. Harrod, K. Hill, J. Hiller, S. Karp, S. Keckler, D. Klein, R. Lucas, M. Richards, A. Scarpelli, S. Scott, A. Snavely, T. Sterling, R. S Williams and K. Yelick, Exascale Computing study: Technology Challenges in Achieving Exascale Systems, Report published under DARPA AFRL contract number FA8650-07-C-7724 (2008).
71. L. W. Liebmann, K. Vaidyanathan and L. Pileggi, *Design Technology Co-Optimization in the era of Sub-Resolution IC Scaling SPIE*, ISBN 9781628419054 (2016).
72. C. Auth, C. Allen, A. Blattner, D. Bergstrom, M. Brazier, M. Bost, M. Buehler, V. Chikarmane, T. Ghani, T. Glassman, R. Grover, W. Han, D. Hanken, M. Hattendorf, P. Hentges, R. Heussner, J. Hicks, D. Ingerly, P. Jain, S. Jaloviar, R. James, D. Jones, J. Jopling, S. Joshi, C. Kenyon, H. Liu, R. McFadden, B. McIntyre, J. Neirynck, C. Parker, L. Pipes, I. Post, S. Pradhan, M. Prince, S. Ramey, T. Reynolds, J. Roesler, J. Sandford, J. Seiple, P. Smith, C. Thomas, D. Towner, T. Troeger, C. Weber, P. Yashar, K. Zawadzki and K. Mistry, A 22nm High Performance and Low-Power CMOS Technology Featuring Fully-Depleted Tri-Gate Transistors, Self-Aligned Contacts and High Density MIM Capacitors, *VLSI Symp. Tech. Dig.*, 131-133 (2012).
73. S. Natarajan, M. Agostinelli, S. Akbar, M. Bost, A. Bowonder, V. Chikarmane, S. Chouksey, A. Dasgupta, K. Fischer, Q. Fu, T. Ghani, M. Giles, S. Govindaraju, R. Grover, W. Han, D. Hanken, E. Haralson, M. Haran, M. Heckscher, R. Heussner, P. Jain, R. James, R. Jhaveri, I. Jin, H. Kam, E. Karl, C. Kenyon, M. Liu, Y. Luo, R. Mehandru, S. Morarka, L. Neiberg, P. Packan, A. Paliwal, C. Parker, P. Patel, R. Patel, C. Pelto, L. Pipes, P. Plekhanov, M. Prince, S. Rajamani, J. Sandford, B. Sell, S. Sivakumar, P. Smith, B. Song, K. Tone, T. Troeger, J. Wiedemer, M. Yang and K. Zhang, A 14nm logic technology featuring 2nd-generation FinFET, air-gapped interconnects, self-aligned double patterning and a 0.0588 μm2 SRAM cell size, *IEEE Inter. Elec. Dev. Meet.*, 3.7.1-3.7.3 (2014).
74. S.-Y. Wu, C. Y. Lin, M. C. Chiang, J. J. Liaw, J. Y. Cheng, S. H. Yang, M. Liang, T. Miyashita, C. H. Tsai, B. C. Hsu, H. Y. Chen, T. Yamamoto, S. Y. Chang, V. S. Chang, C. H. Chang, J. H. Chen, H. F. Chen, K. C. Ting, Y. K. Wu, K. H. Pan, R. F. Tsui, C. H. Yao, P. R. Chang, H. M. Lien, T. L. Lee, H. M. Lee, W. Chang, T. Chang, R. Chen, M. Yeh, C. C. Chen, Y. H. Chiu, Y. H. Chen, H. C. Huang, Y. C Lu, C. W. Chang, M. H. Tsai, C. C. Liu, K. S. Chen, C. C. Kuo, H. T. Lin, S. M. Jang and Y. Ku, A 16nm FinFET CMOS Technology for Mobile SoC and Computing Applications, *IEEE International Elec. Dev. Meet. (IEDM) Tech. Dig.*, 224-227 (2013).
75. K.-I. Seo, B. Haran, D. Gupta, D. Guo, T. Standaert, R. Xie, H. Shang, E. Alptekin, D.-I. Bae, G. Bae, C. Boye, H. Cai, D. Chanemougame, R. Chao, K. Cheng, J. Cho, K. Choi, B. Hamieh,

J. G. Hong, T. Hook, L. Jang, J. Jung, R. Jung, D. Lee, B. Lherron, R. Kambhampati, B. Kim, H. Kim, K. Kim, T. S. Kim, S.-B. Ko, F. L. Lie, D. Liu, H. Mallela, E. Mclellan, S. Mehta, P. Montanini, M. Mottura, J. Nam, S. Nam, F. Nelson, I. Ok, C. Park, Y. Park, A. Paul, C. Prindle, R. Ramachandran, M. Sankarapandian, V. Sardesai, A. Scholze, S.-C Seo, J. Shearer, R. Southwick, R. Sreenivasan, S. Stieg, J. Strane, X. Sun, M. G. Sung, C. Surisetty, G. Tsutsui, N. Tripathi, R. Vega, C. Waskiewicz, M. Weybright, C.-C. Yeh, H. Bu, S. Burns, D. Canaperi, M. Celik, M. Colburn, H. Jagannathan, S. Kanakasabaphthy, W. Kleemeier, L. Liebmann, D. Mcherron, P. Oldiges, V. Paruchuri, T. Spooner, J. Stathis, R. Divakaruni, T. Gow, J. Iacoponi, J. Jenq, R. Sampson and M. Khare, A 10nm Platform Technology for Low Power and High Performance Application Featuring finFET Devices with Multi Workfunction Gate Stack on Bulk and SOI, *VLSI Symposium Technical Digest*, 36-37 (2014).

76. L. T. Clark, V. Vashishthaa, L. Shifrenb, A. Gujjaa, S. Sinhac, B. Clinec, C. Ramamurthya and G. Yericc, ASAP7: A 7-nm finFET predictive process design kit, *Microelectronics J.*, **53** 105-115 (2016).

77. A. Fujimuraa, C. Pierratb, T. Kiuchic, T. Komagatac and Y. Nakagawa, Efficiently writing circular contacts on production reticle, *Proc. SPIE 7748, Symp. photomask and next generation Lithography Mask technology XVII*, 7748 (2010).

78. C. Park, C. Labelle, G. Beique, A. Labonte and D. H. Choi, Challenges of VLSI Patterning and Potential Applications of Atomic Layer Etching, *SEMATECH ALE Workshop* (2014).

79. E. A.-Sanchez, Z Tao, A. Gunay-Demirkol, G. Lorusso, T. Hopf, J.-L. Everaert, W. Clark, V. Constantoudis, D. Sobieski, F. S. Ou and D. Hellin, Self-aligned quadruple patterning to meet requirements for fins with high density, SPIE Newsroom, doi: 10.1117/2.1201604.006378 (2016).

80. R. Xie, A. Knorr, A. Jacob, M. Hargrove, Method of forming fins for FinFET semiconductor devices and selectively removing some of the fins by performing a cyclical fin cutting process, US Patent 9147730 (2015).

81. N. Horiguchi, B. Parvais, T. Chiarella, N. Collaert, A. Veoloso, R. Rooyackers, P. Verheyen, L. Witters, A. Redolfi, A. De Keersgieter, S. Brus, G. Zschaetzsch, M. Ercken, E. Altamirano, S. Locorotondo, M. Demand, M. Jurczak, W. Vanderworst, T. Hoffmann and S. Beisemns, *FinFETs and their Futures in Semiconductor-On-Insulator Materials for Nanoelectronics Applications* 147-148 (Springer-Verlag Berlin Heidelberg 2011).

82. X. Tang, V. Bayot, N. Reckinger, D. Flandre, J.-P. Raskin, E. Dubois and B. Nysten, A Simple Method for Measuring Si-Fin Sidewall Roughness by AFM, *IEEE Trans. Nanotech.*, **8**(5), 611-616 (2009).

83. H. Liu, S. Srivathanakul, H.-W. Liu, S. Gaan, X.-Y. Cai, X.-S. Rao, J. Shu and S. Kim, PMD and STI Gap-Fill Challenges For Advanced Technology of Logic and eNVM, *Elec. Chem. Soc. (ECS) Trans.* **52**(1), 397-402 (2013).

84. F. A. Khaja, H. L. Gossmann, B. Colombeau and T. Thanigaivelan, Bulk FinFET Junction Isolation by Heavy Species and Thermal Implants, *20th International Conference on Ion Implantation Technology (IIT)*, 1~4 (2014).

85. B. S. Wood, F. A. Khaja, B. Colombeau, S. Sun, A. Waite, H. Chen, M. Jin, O. Chan, F. Khaja, T. Thanigaivelan, N. Pradhan, H.- J. Gossmann, S. Sharma, V. R. Chavva, M.-P. Cai, M. Okazaki, S.S. Munnangi, C.-N. Ni, W. Suen, C.-P. Chang, A. Mayur, N. Variam and A. Brand, Fin Doping by Hot Implant for 14nm FinFET Technology and Beyond, *224th Electro Chem. Soc. (ECS) Meet.*, **58**(9), 249-256 (2013).

86. A. P. Jacob, M. K. Akarvardar, J. Fronheiser and W. P. Maszara, Method of forming metastable replacement fins for a finFET semiconductor device by performing replacement growth process, US Patent 20160064250 (2016).

87. A. P. Jacob, M. K. Akarvardar, J. Fronheiser and W. P. Maszara, Methods of forming replacement fins for a FinFET semiconductor device by performing a replacement growth process, US Patent 9240342 (2016).

88. J Fronheiser, M. K. Akarvardar, A. P. Jacob and S Bentley, Method to form defect free replacement fins by H2 anneal, US Patent 9165837 (2015).

89. W.P. Maszara, A. P. Jacob, NV LiCausi, J. A. Fronheiser and K. Akarvardar, Methods of forming FinFET devices with alternative channel materials, US Patent 8580642 (2013).

90. W. P. Maszara, A. P. Jacob, N.V. LiCausi, J. A. Fronheiser and K. Akarvardar, Methods of forming FinFET devices with alternative channel materials, US Patent 8673718 (2014).

91. A. P. Jacob, W. P. Maszara and J. A. Fronheiser, Channel cladding last process flow for forming a channel region on a FinFET device, US Patent 9362405 (2016).

92. Y. Qi, A. P. Jacob, J. A. Fronheiser, M. K Akarvardar and D. P. Brunco, Methods of forming epitaxial semiconductor cladding material on fins of a FinFET semiconductor device, US Patent 14/267634 (2014).

93. R. Xie, P. Montanini, K. Akarvardar, N. Tripathi, B. Haran, S. Johnson, T. Hook, B. Hamieh, D. Corliss, J. Wang, X. Miao, J. Sporre, J. Fronheiser, N. Loubet, M. Sung, S. Sieg, S. Mochizuki, C. Prindle, S. Seo, A. Greene, J. Shearer, A. Labonte, S. Fan, L. Liebmann, R. Chao, A. Arceo, K. Chung, K. Cheon, P. Adusumilli, H.P. Amanapu, Z. Bi, J. Cha, H.-C. Chen, R. Conti, R. Galatage, O. Gluschenkov, V. Kamineni, K. Kim, C. Lee, F. Lie, Z. Liu, S. Mehta, E. Miller, H. Niimi, C. Niu, C. Park, D. Park, M. Raymond, B. Sahu, M. Sankarapandian, S. Siddiqui, R. Southwick, L. Sun, C. Surisetty, S. Tsai, S. Whang, P. Xu, Y. Xu, C. Yeh, P. Zeitzoff, J. Zhang, J. Li, J. Demarest, J. Arnold, D. Canaperi, D. Dunn, N. Felix, D. Gupta, H. Jagannathan, S. Kanakasabapathy, W. Kleemeier, C. Labelle, M. Mottura, P. Oldiges, S. Skordas, T. Standaert, T. Yamashita, M. Colburn, M. Na, V. Paruchuri, S. Lian, R. Divakaruni, T. Gow, S. Lee, A. Knorr, H. Bu and M. Khare, A 7nm FinFET Technology Featuring EUV Patterning and Dual Strained High Mobility Channels, *IEEE International Elec. Dev. Meet. (IEDM) Tech. Dig.*, 2.7.1-2.7.4 (2016).

94. J. Park and C. Hu, Air-Spacer MOSFET With Self-Aligned Contact for Future Dense Memories, *IEEE Elec. Dev. Lett.*, **30**(12), 1368-1370 (2009).

95. S. S. Mujumdar, *Strain Engineering For Strained P-Channel Non-Planar Tri-Gate Field Effect Transistors*, MS Thesis, The Pennsylvania State University (2011).

96. S.-C. Seo, L. F. Edge, S. Kanakasabapathy, M. Frank, A. Inada, L. Adam, M. M. Wang, K. Watanabe, P. Jamison, K. Ariyoshi, M. Sankarapandian, S. Fan, D. Horak, J. T. Li, T. Vo, B. Haran, J. Bruley, M. Hopstaken, S. L. Brown, J. Chang, E. A. Cartier, D.-G. Park, J. H. Stathis, B. Doris, R. Divakaruni, M. Khare, V. Narayanan and V. K. Paruchuri, Full Metal Gate with Borderless Contact for 14 nm and Beyond, *IEEE Symp. Very large Scale Integration (VLSI Symp). Tech. Dig.*, 36-37 (2011).

97. C.-H. Jan, U. Bhattacharya, R. Brain, S.-J. Choi, G. Curello, G. Gupta, W. Hafez, M. Jang, M. Kang, K. Komeyli, T. Leo, N. Nidhi, L. Pan, J. Park, K. Phoa, A. Rahman, C. Staus, H. Tashiro, C. Tsai, P. Vandervoorn, L. Yang, J.-Y. Yeh and P. Bai A 22nm SoC Platform Technology Featuring 3-D Tri-Gate and High-k/Metal Gate, Optimized for Ultra Low Power, High Performance and High Density SoC Applications, *IEEE International Elec. Dev. Meet. (IEDM) Tech. Dig.*, 44-47 (2012).

98. R. Xie, X. Cai, R. Miller and A. Knorr, Methods of forming replacement gate structure for semiconductor devices, US Patent 2013/0187236 (2013).

99. R. Xie, K. Choi, S. C. Fan and S. Ponoth, Methods of forming replacement gate structures for transistors and the resulting devices, US Patent 9257348 (2016).

100. R. Xie, K.-Y. Lim, M. Sung and R. R.-H. Kim, Methods of forming single and double diffusion breaks on integrated circuit products comprised of FinFET devices and the resulting products, US Patent 9412616 (2016).

101. R. Lander, Doping, Contact and Strain Architectures for Highly Scaled FinFETs in CMOS Nanoelectronics edited by N. Collaert, 149-176 (Pan Stanford, Singapore, 2013).
102. L. Pelaz, L. A. Marqués, M. Aboy, P. López, I. Santos and R. Duffy, Atomistic process modeling based on Kinetic Monte Carlo and Molecular Dynamics for optimization of advanced devices, *IEEE Inter. Elec. Dev. Meet.*, 513 (2009).
103. S. Qin, Y. Jeff Hu and A. McTeer, PLAD (Plasma Doping) on 22nm Technology Node and Beyond - Evolutionary and/or Revolutionary, *12th International Workshop on Junction Technology (IWJT)*, 1-11 (2012).
104. S Takeuchi, N. D. Nguyen, F. E. Leys, R. Loo, T. Conard, W. Vandervorst and Matty Caymax, Vapor Phase Doping with N-type Dopant into Silicon by Atmospheric Pressure Chemical Vapor Deposition, *Electro Chemical Soc. Trans.*, **16**(10), 495-502 (2008).
105. K.-W. Ang, J. Barnett, W.-Y. Loh, J. Huang, B.-G. Min, P. Y. Hung, I. Ok, J. H. Yum, G. Bersuker, M. Rodgers, V. Kaushik, S. Gausepohl, C. Hobbs, P. D. Kirsch and R. Jammy, 300mm FinFET results utilizing conformal, damage free, ultra shallow junctions (Xj~5nm) formed with molecular monolayer doping technique, *IEEE International Elec. Dev. Meet. (IEDM) Tech. Dig.*, 837-840 (2011).
106. W. Y. Loh, R. T. P. Lee, R. Tieckelmann, T. Orzali, B. Sapp, C. Hobbs and S.S. Papa Rao, 300mm Wafer Level Sulfur Monolayer Doping for III-V Materials, *SEMI Advanced Semiconductor Manufacturing Conference (ASMC)*, 451-454 (2015).
107. Y.- C. Yeo, Technology Options for Reducing Contact Resistance in Nanoscale Metal-Oxide-Semiconductor Field-Effect Transistors, *5th IEEE Nanoelectronics Conference (INEC)* 128-131 (2013).
108. R. T. P. Lee, A. T.-Y. Koh, W.-W. Fang, K.-M. Tan, A. E.-J. Lim, T.-Y. Liow, C. S.-Yin, A. M. Yong, H. S. Wong, G.-Q. Lo, G. S. Samudra, D.-Z. Chi and Y.-C. Yeo, Novel and cost-efficient single metallic silicide integration solution with dual Schottky-barrier achieved by aluminum inter-diffusion for FinFET CMOS technology with enhanced performance, 2008 IEEE *Symposium on very large scale integration (VLSI) Techn.*, 28-29 (2008).
109. M. Sinha, R. T. P. Lee, S. Nandini Devi, G. Q. Lo, E. F. Chor and Y.-C. Yeo, Single silicide comprising Nickel-Dysprosium alloy for integration in p-and n-FinFETs with independent control of contact resistance by Aluminum implant, 2009 *Symposium on very large scale integration (VLSI) Techn.*, 106-107 (2009).
110. R. T. P. Lee, A. E.-J. Lim, K.-M. Tan, T.-Y. Liow, G.-Q. Lo, G. S. Samudra, D. Z. Chi and Y.-C. Yeo, N-channel FinFETs With 25-nm Gate Length and Schottky-Barrier Source and Drain Featuring Ytterbium Silicide, *IEEE Elect. Dev. Lett.* **28**(2), 164-167 (2007).
111. J. Kedzierski, P. Xuan, E. H. Anderson, J. Bokor, T.-J. King and C. Hu, Complementary silicide source/drain thin-body MOSFETs for the 20 nm gate length regime, *IEEE International Elec. Dev. Meet. (IEDM) Tech. Dig.*, 57-60 (2000).
112. R. T. P. Lee, K.-M. Tan, A. E.-J. Lim, T.-Y. Liow, G. S. Samudra, D. Z. Chi and Y.-C. Yeo, P-Channel Tri-Gate FinFETs Featuring Ni1-yPtySiGeSource/Drain Contacts for Enhanced Drive Current Performance IEEE Elect. Dev. Lett., **29**(5), 438-441 (2008).
113. R. T. P. Lee, K.-M. Tan, T.-Y. Liow, C.-S. Ho, S. Tripathy, G. S. Samudra, D.-Z. Chi and Y.-C. Yeo, Probing the ErSi1.7 Phase Formation by Micro-Raman Spectroscopy, *J. Electrochem. Soc.*, **154**(5), H361-H364 (2007).
114. R. T. P. Lee, T.-Y. Liow, K.-M. Tan, A. E.-J. Lim, H.-S. Wong, P.-C. Lim, D. M. Y. Lai, G.-Q. Lo, C.-H. Tung, G. Samudra, D.-Z. Chi and Y.-C. Yeo, Novel Nickel-Alloy Silicides for Source/Drain Contact Resistance Reduction in N-Channel Multiple-Gate Transistors with Sub-35nm Gate Length, *IEEE International Elec. Dev. Meet. (IEDM) Tech. Dig.*, 1-4 (2006).
115. R. T. P. Lee, T.-Y. Liow, K.-M. Tan, A. E.-J. Lim, C.-S. Ho, K.-M. Hoe, M. Y. Lai, T. Osipowicz, G.-Q. Lo, G. Samudra, D.-Z. Chi and Y.-C. Yeo, Novel Epitaxial Nickel Aluminide-Silicide with Low Schottky-Barrier and Series Resistance for Enhanced

Performance of Dopant-Segregated Source/Drain N-channel MuGFETs, *IEEE Symposium on very large scale integration (VLSI) Technology, Kyoto*, 108-109 (2007).

116. R. T. P. Lee, A. T. Y. Koh, K. M. Tan, T. Y. Liow, D. Z. Chi and Y. C. Yeo, The Role of Carbon and Dysprosium in Ni[Dy]Si:C Contacts for Schottky-Barrier Height Reduction and Application in N-Channel MOSFETs With Si:C Source/Drain Stressors, *IEEE Trans. Elec. Dev.*, **56**(11), 2770-2777 (2009).

117. G. Shine and K. C. Saraswat, Limits of specific contact resistivity to Si, Ge and III-V semiconductors using interfacial layers, *International Conference on Simulation of Semiconductor Processes and Devices (SISPAD)*, 69-72 (2013).

118. D. Connelly, C. Faulkner, P. A. Clifton and D. E. Grupp, Fermi-level depinning for low-barrier Schottky source/drain transistors, *Appl. Phys. Lett.*, **88**, 012105 (2006).

119. J. Y. J. Lin, A. M. Roy, Y. Sun and K. C. Saraswat, Metal-Insulator-Semiconductor Contacts on Ge: Physics and Applications, *2012 International, Meeting on Silicon-Germanium Technology and Device (ISTDM)*, 1-2 (2012).

120. J. Y. Jason Lin, A. M. Roy and K. C. Saraswat, Reduction in Specific Contact Resistivity to Ge Using TiO_2 Interfacial Layer, in *IEEE Elec. Dev. Lett.*, **33**(11), 1541-1543 (2012).

121. J. Robertson and L. Lin, Fermi level pinning in Si, Ge and GaAs systems - MIGS or defects? *IEEE International Elec. Dev. Meet. (IEDM) Tech. Dig.*, 1-4 (2009).

122. H. Yu, M. Schaekers, T. Schram, S. Demuynck, N. Horiguchi, K. Barla, N. Collaert, A. V.-Y. Thean and K. De Meyer, Thermal Stability Concern of Metal-Insulator-Semiconductor Contact: A Case Study of $Ti/TiO_2/n$-Si Contact, *IEEE Trans. Elec. Dev.*, **63**(7), 2671-2676 (2016).

123. J. Borrel, L. Hutin, O. Rozeau, M.-A. Jaud, S. Martinie, M. Gregoire, E. Dubois and M. Vinet, Modeling of Fermi-Level Pinning Alleviation With MIS Contacts: n and pMOSFETs Cointegration Considerations — Part I, in *IEEE Trans. Elec. Dev.*, **63**(9), 3413-3418 (2016).

124. J. Borrel, L. Hutin, O. Rozeau, M.-A. Jaud, S. Martinie, M. Gregoire. E. Dubois and M. Vinet, Modeling of Fermi-Level Pinning Alleviation With MIS Contacts: n and pMOSFETs Cointegration Considerations — Part II, in *IEEE Transactions on Electron Devices*, **63**(9), 3419-3423 (2016).

125. H. Yu, M. Schaekers, S. Demuynck, K. Barla, A. Mocuta, N. Horiguchi, N. Collaert, A. V.-Y. Thean and K. D. Meyer, MIS or MS? Source/drain contact scheme evaluation for 7nm Si CMOS technology and beyond, *16th International Workshop on Junction Technology (IWJT), Shanghai*, 19-24 (2016).

126. K. Majumdar, R. Clark, T. Ngai, K. Tapily, S. Consiglio, E. Bersch, K. Matthews, E. Stinzianni, Y. Trickett, G. Nakamura, C. Wajda, G. Leusink, H. Chong, V. Kaushik, J. Woicik, C. Hobbs and P. Kirsch, Statistical demonstration of silicide-like uniform and ultra-low specific contact resistivity using a metal/high-k/Si stack in a sidewall contact test structure, Symposium on *VLSI Technology (VLSI-Technology): Digest of Technical Papers*, 1-2 (2014).

127. R. T. P. Lee, T-Y. Liow, K.-M. Tan, A. E.-J. Lim, A. T.-Y. Koh, M. Zhu, G.-Q. Lo, G. S. Samudra, D. Z. Chi and Y.-C. Yeo, Achieving Conduction Band-Edge Schottky Barrier Height for Arsenic-Segregated Nickel Aluminide Disilicide and Implementation in FinFETs With Ultra-Narrow Fin Widths, *IEEE Elec. Dev. Lett.*, **29**(4), 382-385 (2008).

128. A. Kaneko, A. Yagishita, K. Yahashi, T. Kubota, M. Omura, K. Matsuo, I. Mizushima, K. Okano, H. Kawasaki, T. Izumida, T. Kanemura, N. Aoki, A. Kinoshita, J. Koga, S. Inaba, K. Ishimaru, Y. Toyoshima, H. Ishiuchi, K. Suguro, K. Eguchi and Y. Tsunashima, High-Performance FinFET with Dopant-Segregated Schottky Source/Drain, *IEEE International Elec. Dev. Meet. (IEDM) Tech. Dig.*, 1-4 (2006).

129. H. S. Wong, L. Chan, G. Samudra and Y. C. Yeo, Sub-0.1-eV Effective Schottky-Barrier Height for NiSi on n-Type Si (100) Using Antimony Segregation, *IEEE Elec. Dev. Lett.*, **28**(8), 703-705 (2007).

130. Y. Matsunaga, S. R. B. Aid, S. Matsumoto, J. Borland and M. Tanjyo, Characterization of BF2, Ga and in dopants in Si for halo implantation, *13th International Workshop on* Junction Technology (IWJT)*, Kyoto*, 74-77(2013).

131. R. T. P. Lee, A. E. J. Lim, K. M. Tan, T. Y. Liow, D. Z. Chi and Y. C. Yeo, Sulfur-Induced PtSi:C/Si:C Schottky Barrier Height Lowering for Realizing N-Channel FinFETs With Reduced External Resistance, *IEEE Elec. Dev. Lett.*, **30**(5), 472-474 (2009).

132. M. Sinha, R. T. P. Lee, S. N. Devi, Guo-Qiang Lo, Eng Fong Chor and Yee-Chia Yeo, p-FinFETs with Al segregated NiSi/p+-Si source/drain contact junction for series resistance reduction, *International Symposium on VLSI Technology, Systems and Applications (VLSI-TSA)* 74-75 (2009).

133. Z. Li, X. An, M. Li, Q. Yun, M. Lin, M. Li, X. Zhang and R. Huang, Morphology and Electrical Performance Improvement of NiGe/Ge Contact by P and Sb Co-implantation, *IEEE Elec. Dev. Lett.*, **34**(5), 596-598 (2013).

134. B. Yang, J.-Y. J. Lin, S. Gupta, A. Roy, S. Liang, W. P. Maszara, Y. Nishi and K. Saraswat, Low-Contact-Resistivity Nickel Germanide Contacts on n+Ge with Phosphorus/Antimony Co-Doping and Schottky Barrier Height Lowering, *International Silicon-Germanium Technology and Device Meeting (ISTDM)*, 1-2 (2012).

135. S. Chen, Y. Wang, C. Heidelberger and M. Thompson, Characterization of dopant diffusion, mobility, activation and deactivation effects for n-type dopants with long-dwell laser spike annealing, *11th International Workshop on Junction Technology (IWJT) Kyoto*, 128-131 (2011).

136. A. T.-Y Koh, R. T. P. Lee, F.-Y Liu, T.-Y. Liow, K. M. Tan, X. Wang, G. S. Samudra, N. Balasubramanian, D.-Z. Chi and Y.-C Yeo, Pulsed Laser Annealing of Silicon-Carbon Source/Drain in MuGFETs for Enhanced Dopant Activation and High Substitutional Carbon Concentration, *IEEE Electron Device Letters*, **29**(5), 464-467 (2008).

137. R. T. P. Lee, A. T.-Y. Koh, F.-Y. Liu, W.-W. Fang, T.-Y. Liow, K.-M. Tan, P.-C. Lim, A. E.-J. Lim, M. Zhu, K.-M. Hoe, C.-H. Tung, G.-Q. Lo, X. Wang, D. K.-Y. Low, G. S. Samudra, D.-Z. Chi and Y.-C. Yeo, Route to Low Parasitic Resistance in MuGFETs with Silicon-Carbon Source/Drain: Integration of Novel Low Barrier Ni(M)Si:C Metal Silicides and Pulsed Laser Annealing, *IEEE International Elec. Dev. Meet. (IEDM) Tech. Dig.*, 685-688 (2007).

138. S. Zhou, F. Liu, S. Prucnal, K. Gao, M. Khalid, C. Baehtz, M. Posselt, W. Skorupa and M. Helm, Hyperdoping silicon with selenium: solid vs liquid phase epitaxy, *Nature Scientific Reports*, **5**, 8329 (2015).

139. A. Dimoulas, P. Tsipas, A. Sotiropoulos and E. K. Evangelou, Fermi-level pinning and charge neutrality level in germanium, *Appl. Phys. Lett.*, **89**, 252110 (2006).

140. T. Nishimura, K. Kita and A. Toriumi, Evidence for strong Fermi-level pinning due to metal-induced gap states at metal/germanium interface, *Appl. Phys. Lett.*, **91**, 123123 (2007).

141. R. T. P. Lee, D. Z. Chi and Y. C. Yeo, Platinum Germanosilicide as Source/Drain Contacts in P-Channel Fin Field-Effect Transistors (FinFETs), in *IEEE Trans. on Elec. Dev.*, **56**(7), 1458-1465 (2009).

142. S. Zhu, R. Li, S. J. Lee, M. F. Li, A. Du, J. Singh, C. Zhu, A. Chin and D. L. Kwong, Germanium pMOSFETs with Schottky-barrier germanide S/D, high-κ gate dielectric and metal gate, in *IEEE Electron Device Letters*, **26**(2), 81-83 (2005).

143. Rui Li, S. J. Lee, H. B. Yao, D. Z. Chi, M. B. Yu and D. L. Kwong, Pt-Germanide Schottky source/drain Germanium p-MOSFET with HfO2 gate dielectric and TaN gate electrode, in *IEEE Electron Device Letters*, **27**(6), 476-478 (2006).

144. C. Chui and K. C. Saraswat, Nanoscale Germanium MOS Dielectrics and Junctions *in Germanium-Based Technologies from Materials to Devices Edited by C. Claeys and E. Simoen*, 337-356 (Elsevier, Amsterdam 2007)

145. O. Nakatsuka, A. Suzuki, A. Sakai, M. Ogawa and S. Zaima, Impact of Pt Incorporation on Thermal Stability of NiGe Layers on Ge(001) Substrates, *2007 International Workshop on Junction Technology, Kyoto*, 87-88 (2007).

146. H.-S. Shin, S.-K. Oh, M.-H. Kang, J.-H. Jang, J. Oh, P. Majhi, R. Jammy, Y.-S. Chung, S.-S. Kim, D.-S. Lee, S.-J. Lee and H.-D. Lee, Improvement of thermal stability of Ni-germanide with co-sputtering of nickel and palladium for high performance Ge CMOSFET, *International Symposium on Semiconductor Device Research (ISDRS)*, 1-2 (2011).

147. J.-W. Lee, J.-H. Bae, J.-H. Hwang, H.-K. Kim, M.-H. Park, H. Kim and C.-W. Yang, Dynamic study on microstructural evolution of nickel germanide utilizing zirconium interlayer, *Microelectronic Eng.*, **89**, 23-26 (2012).

148. S. L. Liew, R. T. P. Lee, K. Y. Lee, B. Balakrisnan, S. Y. Chow, M.Y. Lai and D.Z. Chi, Enhanced morphological stability of NiGe films formed using Ni(Zr) alloy, *Thin Solid Films*, **504**(1-2), 104-107 (2006).

149. R. T. P. Lee, W. Y. Loh, R. Tieckelmann, T. Orzali, C. Huffman, A. Vert, G. Huang, M. Kelman, Z. Karim, C. Hobbs, R. J. W. Hill and S. S. P. Rao, Technology Options to Reduce Contact Resistance in Nanoscale III-V MOSFETs, Electro Chemical Society *(ECS) Transactions* **66**(4), 125-134 (2015).

150. M. Abraham, S.-Y. Yu, W. H. Choi, R. T. P. Lee and S. E Mohney, Very low resistance alloyed Ni-based ohmic contacts to InP-capped and uncapped n+-In0. 53Ga0.47As, *J. App. Phys.*, **116**(16), 164506 (2014)

151. L. A. Walsh, G. Hughes, C. Weiland, J. C. Woicik, R. T. P. Lee, W.-Y. Loh, P. Lysaght and C. Hobbs, Ni-(In, Ga) As alloy formation investigated by hard-X-ray photoelectron spectroscopy and X-ray absorption spectroscopy, *Phys. Rev. Applied*, **2**(6), 064010 (2014).

152. L. A. Walsh, C. Weiland, A. P. McCoy, J. C. Woicik, R. T. P. Lee, P. Lysaght and G. Hughes, Investigation of the thermal stability of Mo-In0. 45Ga0. 47As for applications as source/drain contacts, *J. of App. Phys.*, **120**(13), 135303 (2016).

153. S. K. Tiku and W. M. Duncan, Self-Compensation in Rapid Thermal Annealed Silicon-Implanted Gallium Arsenide, *J. Electrochem. Soc.*, **132**, 2237 (1985).

154. R. Venkatasubramanian, K. Patel and S. K. Ghandhi, Compensation mechanisms in n+-GaAs doped with silicon, *J. Crystal Growth*, **94**, 34 (1989).

155. J. P. Donnelly, W.T. Lindley and C.E. Hurwitz, Silicon-and selenium-ion-implanted GaAs reproducibly annealed at temperatures up to 950° C, *Appl. Phys. Lett.*, **27**, 41 (1975).

156. T. Orzali, A. Vert, R. T. P. Lee, A. Norvilas, G. Huang, J. L. Herman, R. J.W. Hill and S. S. Papa Rao, Heavily tellurium doped n-type InGaAs grown by MOCVD on 300 mm Si wafers, *J. Crystal Growth*, **426**, 243-247 (2015).

157. C. J. Pinzone, N.D. Gerrard, R. D. Dupuis, N.T. Ha and H. S. Luftman, Heavily-doped n-type InP and InGaAs grown by metalorganic chemical vapor deposition using tetraethyltin, *J. Appl. Phys.*, **67**, 6823 (1990).

158. R. T. P. Lee, Y. Ohsawa, C. Huffman, Y. Trickett, G. Nakamura, C. Hatem, K. V. Rao, F. Khaja, R. Lin, K. Matthews, K. Dunn, A. Jensen, T. Karpowicz, Peter F. Nielsen, E. Stinzianni, A. Cordes, P. Y. Hung, D.-H. Kim, R. J. W. Hill, W.-Y. Loh, C. Hobbs, "Ultra low contact resistivity (< $1 \times 10-8$ Ω-cm2) to $In_{0.53}Ga_{0.47}As$ fin sidewall (110)/(100) surfaces: Realized with a VLSI processed III-V fin TLM structure fabricated with III-V on Si substrates, *IEEE International Electron Devices Meeting (IEDM) Tech. Dig.*, 32.4.1-32.4.4. (2014).

159. H. Kawasaki, V. S. Basker, T. Yamashita, C.-H. Lin, Y. Zhu, J. Faltermeier, S. Schmitz, J. Cummings, S. Kanakasabapathy, H. Adhikari, H. Jagannathan, A. Kumar, K. Maitra, J. Wang, C.-C. Yeh, C. Wang, M. Khater, M. Guillorn, N. Fuller, J. Chang, L. Chang, R. Muralidhar, A. Yagishita, R. Miller, Q. Ouyang, Y. Zhang, V. K. Paruchuri, H. Bu, B. Doris, M. Takayanagi, W. Haensch, D. McHerron, J. O'Neill and K. Ishimaru, Challenges and solutions of FinFET

integration in an SRAM cell and a logic circuit for 22 nm node and beyond, *IEEE International Elec. Dev. Meet. (IEDM) Tech. Dig.*, 1-4 (2009).

160. R. Xie, C.-C. Yeh, X. Cai and Q. Liu., Series resistance reduction in vertically stacked silicon nanowire transistors, US Patent 14/739543 (2015).

161. R. Xie and A. Knorr, Integrated circuit product comprising lateral and vertical FinFet devices, US Patent 9443976 (2016).

162. A. V. Y. Thean, D. Yakimets, T. Huynh Bao, P. Schuddinck, S. Sakhare, M. G. Bardon, A. Sibaja-Hern,ez, I. Ciofi, G. Eneman, A. Veloso, J. Ryckaert, P. Raghavan, A. Mercha, A. Mocuta, Z. Tokei, D. Verkest, P. Wambacq, K. De Meyer and N. Collaert, Vertical Device Architecture for 5nm, beyond: Device & Circuit Implications, *VLSI Symposium Technical Digest*, T26-T27 (2015).

163. R. Xie and A. Knorr, Methods of forming lateral, vertical FinFET devices, the resulting product, US Patent 9245885 (2016).

164. P. C. Andricacos, Copper on-chip interconnections, a breakthrough in electrodeposition to make better chips, *Electrochemical Society Interface*, **8**(1) 32-37 (1999).

165. R. Venkatraman, E Weitzman and R. Fiordalice, Method of forming an interconnect structure, US Patent 5814557 A (1998).

166 H. Shimizu, K. Shima, Y. Suzuki, T. Momosea and Y. Shimogakiaet, Precursor-based designs of nano-structures, their processing for Co(W) alloy films as a single layered barrier/liner layer in future Cu-interconnect, *J. Mater. Chem. C*, **3**(11) 2500-2510 (2015).

167. R. Mehta, S. Chugh and Z. Chen, Enhanced electrical, thermal conduction in graphene-encapsulated Cu nanowires, *Nano Letters*, **15**(3) 2024-2030 (2015).

168. A. Grill, PECVD low, ultralow dielectric constant materials: From invention, research to products, *J. Vac. Sci., Tech.* B **34**, 020801(2016).

169. S. Y. Fang, Y.-S. Tai and Y.-W. Chang, Layout decomposition for Spacer-is-Metal (SIM) self-aligned double patterning, *The 20th Asia, South Pacific Design Automation Conference, Chiba*, 671-676 (2015).

170. R.-H. Kim, C.-S. Koay, S. D. Burns, Y. Yin, J. C. Arnold, C. Waskiewicz, S. Mehta, M. Burkhardt, M. E. Colburn and H. J. Levinson, Spacer-defined double patterning for 20nm, beyond logic BEOL technology, *Proceedings SPIE*, 7973, 79730N (Optical Microlithography XXIV, 2011).

High-Speed SiGe BiCMOS Technologies and Circuits

A. Mai[*], I. Garcia Lopez, P. Rito, R. Nagulapalli, A. Awny, M. Elkhouly, M. Eissa,
M. Ko, A. Malignaggi, M. Kucharski, H. J. Ng, K. Schmalz and D. Kissinger

*IHP, im Technologiepark 25,
15236 Frankfurt (Oder), Germany*
mai@ihp-microelectronics.com

This work reports on the development of SiGe-BiCMOS technologies for mm-wave and THz high frequency applications. We present state-of-the-art performances for different SiGe heterojunction bipolar transistor (SiGe-HBT) developments as well as the evolution of complex BiCMOS technologies. With respect to different technology generations of high-speed SiGe-BiCMOS processes at IHP we discuss selected device modifications of the SiGe-HBT to achieve high frequency performances of a complex BiCMOS technology towards the 0.5 THz regime. We show the difference of high-frequency performance difference with respect to maximum achievable transit frequencies f_T and oscillation frequencies f_{max} in comparison to RF-CMOS technologies and depict the required increase of additional process effort for the HBT-module integration for a 0.5 THz SiGe-BiCMOS technology. Moreover different high speed circuits are presented like broadband ICs for optical communication, high frequency circuits for wireless communication at 60 and 240 GHz, mm-wave radar circuits at 60 and 120 GHz as well as THz circuits operating at 245 GHz and 500 GHz for spectroscopic applications. All reviewed circuit examples are based on the discussed 130nm-SiGe-BiCMOS technologies and show their potential for a broad range of high-speed applications.

Keywords: SiGe-BiCMOS; mm-wave and THz: broadband and wireless communication, radar circuits.

1. Introduction

The evolution of the semiconductor industry and in particular the silicon based microelectronics in the past decades changed almost every part of our life. Academic and industrial research, actually every job relation and especially our social life with respect to our way of communication changed dramatically. Today our world is wireless. Mobile and local area networks are accessible at almost every place and every time by smartphones, tablets and other consumer electronics. Whereas microprocessors are based on digital electronics and therefor CMOS (complementary-metal-oxide-semiconductor) technologies the components for the interaction between different devices is often related to analog-mixed-signal (AMS) circuits. Almost thirty years ago the first silicon-

---

[*]Corresponding author.

germanium heterojunction bipolar transistor (SiGe-HBT) was demonstrated[1,2] based on molecular-beam-epitaxial SiGe base process and only few years later a SiGe HBT with cutoff frequencies f_T of 75 GHz was presented and showed the potential for high frequency applications by exceeding the performance of standard Si bipolar junction transistors (Si-BJT) by an factor of two[4]. Improvements of the devices and process technology grew rapidly at this time and after first announcement of a SiGe-BiCMOS technology[5] in 1992 large-scale integrated circuits were realized like digital-analog converters[6] and SiGe-BiCMOS technologies were widely used in AMS-circuits. Main advantages of SiGe-HBTs are the obvious better radio-frequency (RF) performance and certain properties as low-noise in comparison to standard silicon bipolar and CMOS devices. However, the development of RF-CMOS technologies moved on and certain markets like the 2.4 GHz WLAN is nowadays dominated by these technologies.

But novel broadband data communication standards require even devices with high-performance, higher operating frequencies and moreover integrated circuits with ever increasing complexity. To enable greater chip integration with reduced power consumption and optimized costs silicon based technologies are essential. Nowadays novel high speed and mm-wave applications as optical networks, automotive radar or imaging systems, local area networks (LAN) RF transceivers for 60GHz operation or Point-to-Point radio applications increased the demand for new generations of silicon based high frequency technologies especially in the last years.

In this paper the authors review performance status of SiGe-BiCMOS technologies. We present recent device developments with respect to different generations of SiGe-BiCMOS technologies at IHP (Innovations for high performance microelectronics). In chapter III we review different SiGe-HBTs in a 0.25μm as well as 0.13μm baseline CMOS technologies and consider certain device optimization for the decrease of parasitics and therefor improvement of the high frequency performance towards 0.5 THz. Chapter IV to VII present different integrated mm-wave circuits like broadband ICs for optical communication, high frequency circuits for wireless communication at 60 and 240 GHz, mm-wave radar circuits at 60 and 120 GHz as well as THz circuits operating at 245 GHz and 500 GHz for spectroscopic applications. All circuit examples are based on the discussed 130nm-SiGe-BiCMOS technologies of IHP.

2. State of the Art SiGe-HBT and BiCMOS Performance

The high frequency performance in particular the increase of f_T as well as the maximum oscillation frequency f_{max} of SiGe-HBTs was pushed and exceeded the 100 GHz regime in the early nineties[7,8]. With the first available 200mm wafer process SiGe BiCMOS technology[9] in 1994 the technologies could fulfill upcoming market demands and several companies and institutes developed SiGe BiCMOS technologies with an increasing performance[10-14]. At this time a wide range of different market segments in wireless and cellular applications for frequencies of 900MHz or 2.4 GHz in the field of wireless local area networks (WLAN)[15,16] were covered by SiGe-BiCMOS technologies and prove their usefulness especially in the high frequency market[17]. Due to the improvements of RF-

CMOS performance first circuit demonstrations predicted the replacement of SiGe chips in these markets[18]. However, for other applications like transceiver circuits for optical networks or pre-amplifiers with large bandwidths[19,20] SiGe-BiCMOS technologies were still used. In the last years the boom of broadband communication applications increased the demand for high performance devices and technologies which offer increased circuit complexity and exceed the limits of available RF CMOS technologies. As shown in Fig. 1 the performance of SiGe HBTs and their integration in BiCMOS technologies were continuously pushed forward and first devices and technologies exceeding the 200 GHz regime could be demonstrated[21,22] in the last decade. However, technologies with cut-off frequencies f_T and f_{max} which simultaneously exceed 300 GHz and 400 GHz couldn't be realized until 2010. In the European Union funded Dotfive project (2008-2011) a SiGe-HBT was presented with f_{max}/f_T of 500GHz/300GHz for the first time[23] and finally transferred into a full SiGe-BiCMOS technology platform[24] for mm-wave and THz circuits.

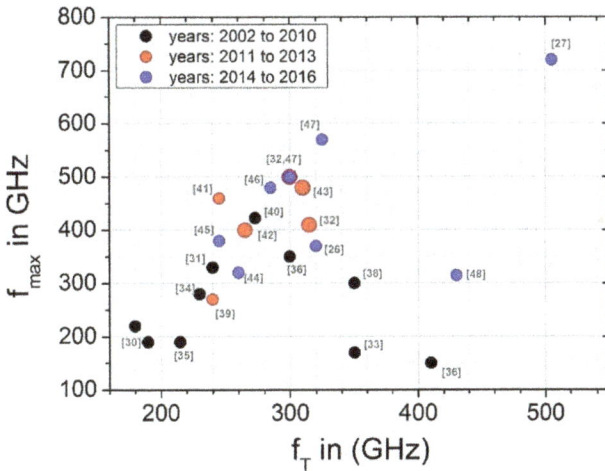

Fig. 1. Peak f_T vs. f_{max} data of selected SiGe-HBT technologies and discrete device developments. Different colors indicate various time periods of the first announcement of these technologies.

Among other things these results encouraged new developments and a continuous progress of technology improvements in industrial processes[25,26] could be observed. However, no SiGe-HBT technology or device development could exceed a certain frequency regime. Either SiGe-HBTs with high f_{max} above 400 GHz and at f_T below 350 GHz or with high f_T values above 350 GHz at simultaneously lower f_{max} below 300 GHz could be demonstrated. These limits could be recently broke through with the completion of the EU-Dotseven project in 2016. Here a SiGe-HBT with f_T/f_{max} of 505 GHz/720 GHz[27] could be presented.

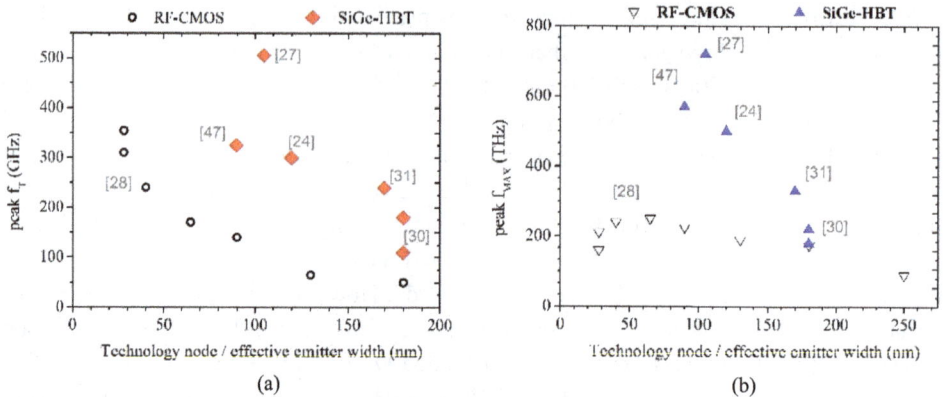

Fig. 2. Peak cut-off frequencies f_{max} (a) and f_T (b) for different SiGe-HBTs and a RF-CMOS technology vs. the effective emitter width and technology node, respectively.

The integration of SiGe devices with very high frequency performance in CMOS environments combines the advantages of two technologies in a single chip. In comparison to RF-CMOS technologies SiGe-HBTs offer at least a two to three times better frequency performance at the same technology node (Fig. 2). The peak f_T value of RF-CMOS devices benefits obviously from the continuous scaling of the technology node[28]. Certain device and process adaptations of the SiGe-HBT which will be described in chapter 3 show that even for an effective emitter width of 170nm f_T could be doubled. Hence a scaling of the HBT devices is not as crucial to get an obvious high f_T and f_{max} increase as for RF-CMOS transistors. Nevertheless, the SiGe-HBT devices realized in a 130nm BiCMOS process benefits among other things from a reduced emitter window width and show maximum f_T values within the range or better than a 28nm RF-CMOS process. Moreover Fig. 2b shows the behavior of f_{max} for SiGe HBTs and RF-CMOS transistors in dependence of the effective emitter width and the corresponding CMOS technology node, respectively[29].

3. Device and Technology Developments

This chapter describes device optimizations and integration concepts for SiGe-HBTs in baseline 250nm and 130nm CMOS technologies at IHP. Different device concepts are compared to show the progress for the integration and the increase of process effort for the fabrication of high performance HBT modules. The last part of the chapter reviews certain device optimizations to enable SiGe-HBTs with frequencies in the 0.5 THz range and summarizes some performance characteristics.

3.1. *SiGe-HBT module and device developments for 0.5 THz operation*

Different high speed SiGe-BiCMOS technology from IHP based on a 0.25μm CMOS platform and with HBT cutoff frequencies f_T and f_{max} up to 200 GHz[30] rely in particular on a novel low-cost, epitaxial-free collector design. Due to adaptations of the integration concept by using an elevated external base region and changed germanium and boron profiles in the SiGe-base an improvement of the performance up to f_T/f_{max} of 240/330 GHz could be achieved in a novel 130nm SiGe-BiCMOS technology[31]. This RF-BiCMOS technology offers dual gate CMOS devices with 2nm and 7nm gate-oxides for 1,2V-digital and 3.3-IO transistors, respectively. A silicidation blocking mask is used to realize a set of different poly-resistors. The back-end-of-line (BEOL) consists of five thin aluminum metal layers including a MIM-capacitor and two thick metal layers of 2μm and 3μm thickness for the fabrication of antennas and high-Q inductors.

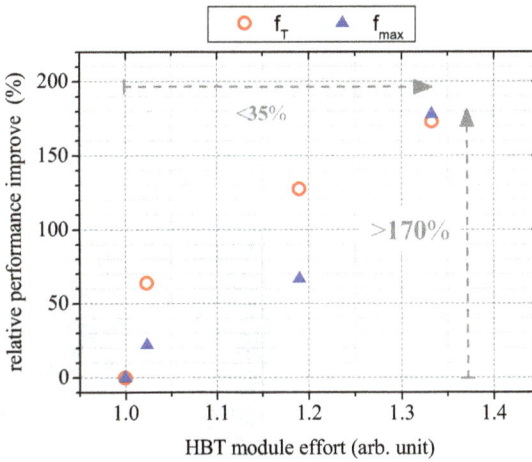

Fig. 3. Relative increase of cut-off frequencies f_T and f_{max} for different HBT generations and SiGe–BiCMOS technologies of IHP versus increase of HBT module integration effort[29].

Figure 3 shows the relative RF-performance increase of different HBT generations with respect to the required additional process effort. We estimate the necessary effort for the integration of the different HBT by summarizing different process steps of the HBT module integration to certain step clusters[29]. Despite to the fact that the integration became more challenging the increase of additional process steps is only 35% and leads to a tremendous improvement of the SiGe-BiCMOS performance by 170%. This was achieved in particular by optimization of single process steps of the HBT module integration e.g. SiGe-base epitaxy. Figure 4 shows a schematic cross section (4a) of a half-terahertz SiGe-HBT and a transmission-electron-microscope image of the realized device. Certain device modifications for the fabrication of IHP's SG13S technology with f_T/f_{max} of 240/300 GHz enabled a performance improve towards f_T/f_{max} of 300/500 GHz in IHP's SG13G2 technology[32]. Both concepts use the same integration concept and baseline CMOS process. Moreover they have an elevated base enhancement of the poly-silicon base contact regions in common and a self-aligned separation by a spacer d_{sp} between emitter window and the heavily doped base link region. The transit frequency f_T

benefits from short transit times τ_F, high collector current densities I_C and therefor short charging times of the base-emitter C_{BE} and base-collector C_{BC} capacitances as well as reduced emitter- and collector resistances R_E and R_C. Reduced τ_F and high I_C require steep vertical doping profiles in the SiGe base. Beside high f_T a low base resistance R_B is required to achieve high f_{max}. Several modifications like

- source/drain spike anneal
- substrate orientation
- base profile adaption
- d_{sp} spacer reduction

were introduced and require process adaptations but no additional steps[32]. For this reason the total increase of additional process steps for the SiGe HBT module of SG13G2 in comparison to the first generation SG13S is only approximately 10%. Process modifications which require additional process steps are

- Shrinking of the emitter window width W_E
- Adapted Selectively-Implanted-Collector (SIC) formation.

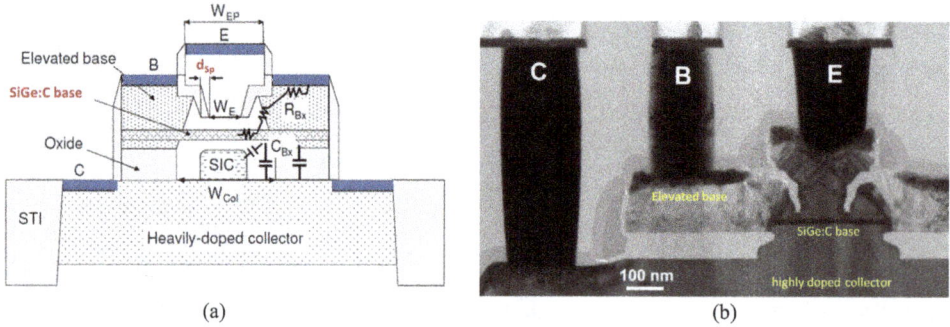

Fig. 4. Schematic (a) and TEM cross section (b) of 0.5 THz SiGe HBT. Key device dimensions are depicted in (a)[31].

3.2. *SiGe-HBT device characteristics*

The device characteristic presented in Fig. 5 and Fig. 6 were measured for HBTs of different technologies and various emitter window dimension of $w_E = 120$nm for SG13S and SG13G2 and 170nm for H1-technology. The transit frequency f_T and the maximum oscillation frequency f_{max} were extracted from s-parameter measurements extrapolating current gain h_{21} and unilateral gain U in the frequency band of 30 to 50 GHz. Figure 5 shows the f_T and f_{max} versus the collector current density. For f_{max} values of 330 GHz the 0.5 THz HBT of SG13G2 requires only 20% of the collector current density compared to the device of the former technology generation SG13S. With respect to the peak f_{max} value of almost 200 GHz for the SG25H1-technology we obtain only 10% of the necessary collector current density of the SG13G2 HBT. Finally it corresponds to obviously decreased power consumption for the new technology generations and frequencies below 300 GHz.

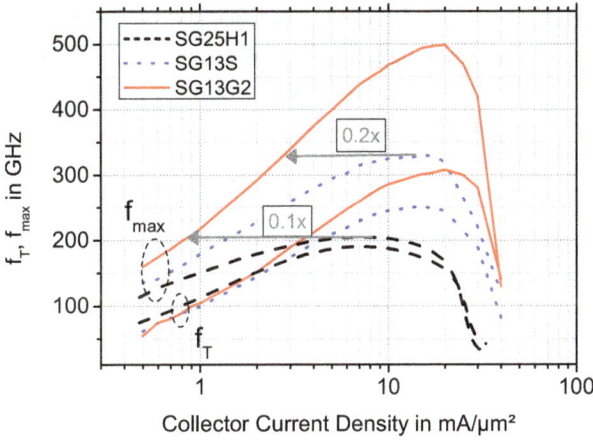

Fig. 5. Transit frequency f_T and maximum oscillation frequency f_{max} versus collector current density for three different SiGe-HBT generations of IHPs BiCMOS technologies.

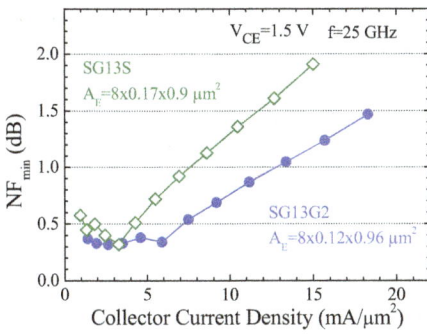

(a)

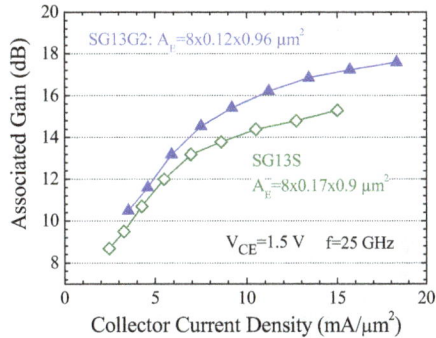

(b)

Fig. 6. SG13S and SG13G2 minimum noise figure NF_{min} at 25 GHz (a) and the associated gain as a function of the collector current density[32].

Figure 6 shows the minimum noise figure and the associated gain at 25 GHz. Moreover an increased f_T and reduced base resistance R_B improves the noise figure and we obtain a lower NF_{min} (Fig. 6a) for the 0.5 THz HBT of SG13G2. Data points of Fig. 6b were measured at V_{BE} step variation of 0.1 V and the values for the maximum collector current density corresponds to peak f_T at $V_{BE} = 0.92 V$[32].

4. Broadband High Speed ICs for Optical Communications

The advantages of SiGe BiCMOS technology in terms of excellent noise performance, high speed (high f_T and f_{max}), high transconductance and the ability to integrate digital functionalities through the CMOS module, make it a very attractive candidate for the implementation of integrated circuits for optical front-ends. In the following, some examples of state-of-the-art transimpedance amplifiers (TIAs) and Mach-Zehnder modulator (MZM) drivers implementations, which benefit from the above mentioned advantages of the SiGe BiCMOS technologies, are described.

4.1. *Optical transmitter: Driver circuits for Mach-Zehnder modulators*

A hybrid optical transmitter module comprising a 15-segment InP IQ segmented Mach-Zehnder modulator (IQ-SEMZM) and two SiGe:C BiCMOS drivers fabricated in the 0.13μm technology from IHP, featuring f_T / f_{max} = 250/340 GHz and BV_{CEO} = 1.7 V, has been implemented[49,50]. The hybrid configuration enables each component to be custom-tailored, combining the assets of the SiGe HBTs, able to provide larger output swings than high speed CMOS digital drivers at similar f_T, improving optimization possibilities, along with the higher efficiency and smaller chip size of the InP MZMs compared to depletion-type SiPh and $LiNbO_3$ MZMs. Fig. 7a shows the photo fo the packaged module and Fig. 7b presents a detailed view of the IQ-SEMZM and BiCMOS drivers bonding.

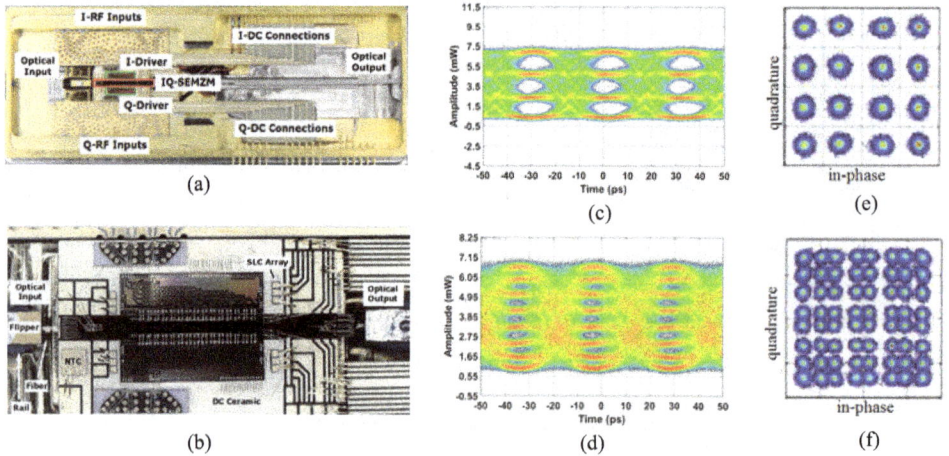

Fig. 7. Picture of the packaged module with the IQ-SEMZM marked in red and the driver ICs in green (a), detailed view of IQ-SEMZM and BiCMOS drivers bonding (b), 4-PAM (c) and 8-PAM (d) at 32 GBd and IQ constellations of the PDM 16-QAM and PDM 64-QAM (f) at maximum optical-signal-to-noise-ratio (OSNR) at 32 GBd.

The IQ drivers, occupying an area of 7.5 mm² each, feature integrated 4-bit DAC functionality and deliver a maximum differential output voltage of 2.5 Vpp across all 15 segments, which well satisfies the $2V\pi$ requirements of the modulator. Each driver dissipates ~1 W of power, when delivering the maximum output voltage. The integrated 4-bit DAC functionality is implemented by means of a wired binary-to-thermometer encoder: four single-ended 50 Ω microstrip lines are adequately routed, distributing the input RF signals to a thermocode-weighted amount of driver cores. The length of the microstrip lines between segments is designed such that the electrical delay matches the optical delay per segment of 3.7 ps. The module allows a wide range of modulation formats, among which up to 8-PAM and polarization-division-multiplexed (PDM) 64-QAM back-to-back (B2B) error-free electro-optical (E/O) transmission at record speed of 32 GBd have been demonstrated. The 32 GBd PDM 64-QAM signal was transmitted error-free over 80 km of standard-single-mode-fiber (SSMF), featuring 7.8 pJ/bit record low energy consumption. Fig. 7c and 7d show the 4-PAM and 8-PAM eye diagrams at 32 GBd, respectively. The 16-QAM and 64-QAM constellation diagrams for the single polarization corresponding to the B2B configuration at 32 GBd are presented in Fig. 7e and Fig. 7f respectively. The proposed arrangement, featuring the highest reported modulation format at this data rate, without compromising on power dissipation, proves the suitability of the SiGe HBT driving concept to bring current transmitter modules ahead of the state-of-the-art.

Not only is the SiGe BiCMOS process a good candidate for implementation of integrated circuits for optical front-ends using hybrid integration with III-V technologies, but also for monolithically-integrated technologies, in which the photonic as well as the electronic components are fabricated together on the same die, reducing interconnection parasitics between devices. Another advantage of this monolithically integrated Si-photonics technology is the exploitation of the excellent yield of the matured Si technologies and the high-volume in production because of the wafer-sizes.

In light of this, a monolithically integrated driver and MZM is presented[51,52], comprising for the first time linear driver and segmented depletion-type MZM in a single chip using the electronic-photonic integrated circuit (EPIC) 0.25 μm SiGe:C BiCMOS technology platform from IHP. The segmented approach, used in the hybrid implementation as well, is here a main requisite for the driver topology to be able to deliver the required voltage to the phase shifters of the modulator of 6.08 mm long. The driver incorporates linear amplifiers, allowing the usage of any analogue signal for the transmission, essential for future coherent high-order modulation communications (such as PAM and QAM). Moreover, it can also be used to apply pre-emphasis to correct the nonlinear transfer function of the MZM.

The block diagram of the transmitter is presented in Fig. 8a. The two phase shifters of the MZM are divided into 16 segments, electrically approximated to a RC lumped model. The driver integrates two stages: an input stage which matches the RF input signal drives the electrical transmission line (terminated by 40 Ω resistors) and gives the bias for the next stage; and a second stage which is distributed in the same number of segments laterally to the MZM, applying the required voltage amplitude of 4 Vppd. The block diagram can be compared to the chip microphotograph, presented in Fig. 8b. The monolithically integrated linear driver and modulator occupy an area of 12.7 mm^2 (9.8 mm x 1.3 mm).

(a)

(b)

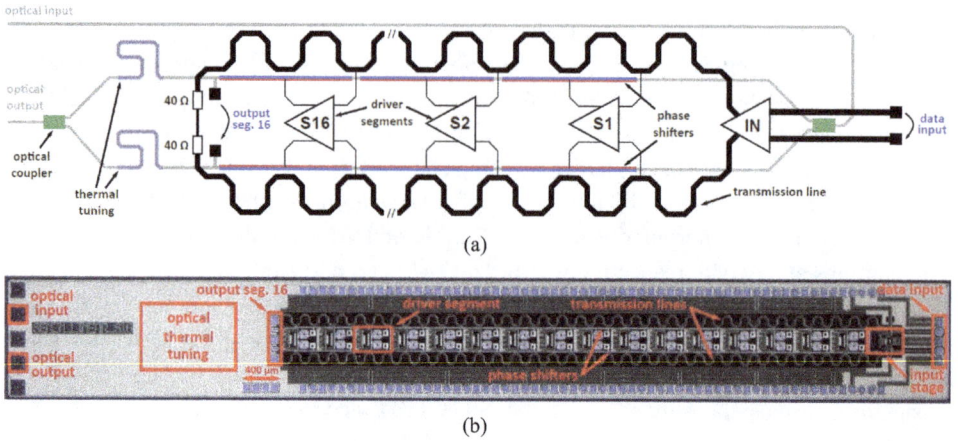

Fig. 8. Block diagram of the monolithically integrated segmented driver and MZM (a), chip microphotograph of the optical transmitter (b).

The total gain of the driver is 13 dB and features a total harmonic distortion (THD) below 5%. Electro-optical time-domain measurements using PAM-4 modulation format were performed, demonstrating optical eye-diagrams up to 25 Gbaud. The electro-optical bandwidth of the transmitter is 18 GHz[51]. The power dissipation of the driver is 1.5 W, resulting in an energy per bit of 30 pJ/bit at 50 Gb/s. Electro-optical measurements with OOK signal were also performed when increased the DC power of the driver up to 2 W to show a maximum extinction ratio (ER) of 13 dB at 28 Gb/s [52].

4.2. Optical receiver: Transimpedance amplifiers

Figure 9 shows a differential TIA employed in a coherent detection receiver system[53]. The TIA consists of two independent input transimpedance stages to process the input signals coming from two balanced photodiodes, two variable gain amplifiers (VGAs) and an output buffer to drive 100 Ω differential load. It operates in two modes: the manual gain control mode (MGC) in which the gain is controlled manually by means of the GC pin and the automatic gain control mode (AGC) in which a control loop regulates automatically the gain of the two VGAs to maintain the output amplitude at a specified level (controlled by the pin OA), when the input amplitude changes. In this design the replica TIA is used for three purposes: cancellation of the input DC-overload current, cancellation of the DC-offset between the two input currents from the balanced photodiodes due to manufacturing tolerances and also for the stabilization of the biasing currents of the two input transimpedance shunt-feedback stages against changes in the power supply. The TIA achieves 53 GHz of 3-dB bandwidth, suitable for operation at 64 GBd. Measured low-frequency transimpedance is 80 dBΩ. The TIA provides 900 mVpp to 100 Ω differential loads with THD below 5% for input differential currents up to 3 mApp. The input-referred noise spectrum density is calculated to be around 25 pA/sqrt(Hz). The chip is implemented in IHP SG13S SiGe BiCMOS technology with

$f_T/f_{max} = 250$ GHz/340 GHz. It occupies 2 mm^2 and contains two TIAs to facilitate coherent detection with I and Q signals. The power dissipation per TIA is 277 mW. Hybridly-integrated with the photodiode and 90° optical-hybrids (fabricated in III-V technology), the coherent optical-electrical frontend has been used to transmit 128 Gb/s using QPSK modulation with 64 GBd.

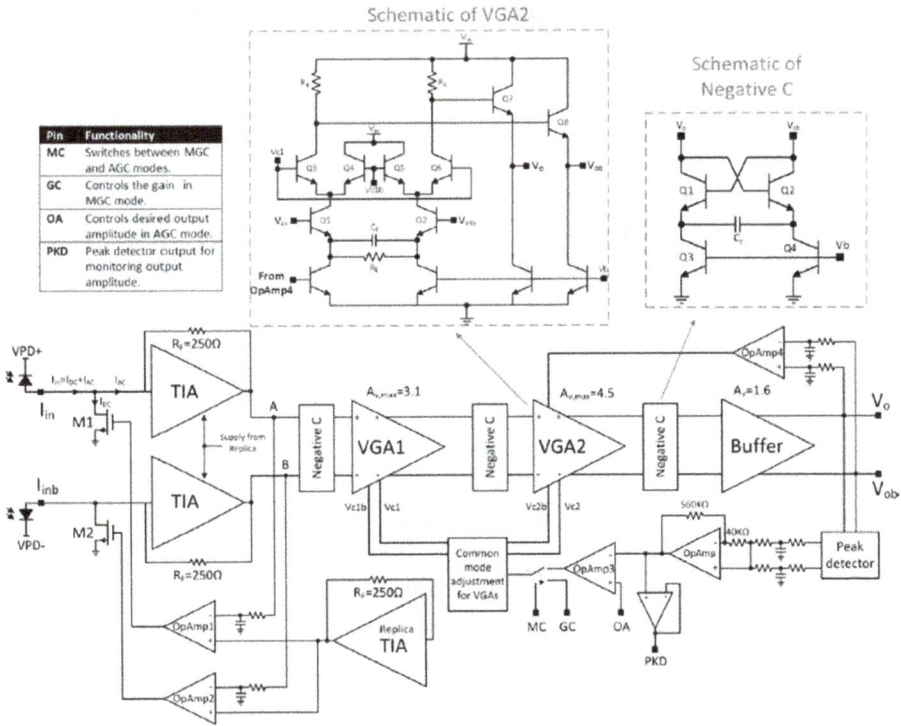

Fig. 9. Block diagram of the TIA–AGC.

Fig. 10. Integrated opto-electronic receiver microphotograph.

As in the transmitter case, a monolithically-integrated linear receiver for direct detection has been as well demonstrated[54,55]; Fig. 10 shows the block diagram. The receiver incorporates a Ge photodiode with a bandwidth > 40 GHz and responsivity of 0.84 A/W at a reverse bias of -2 V. The integrated receiver achieves an optical/electrical bandwidth of 36 GHz, while having 66 dBΩ of transimpedance gain. The equivalent input noise spectrum density is found to be 22.3 pA/sqrt(Hz). The receiver achieves a BER of 10^{-12} and 10^{-7} for 40 and 56 Gb/s, respectively in on-off-keying modulation.

5. High-frequency Integrated Circuits for Wireless Communication and Sensing

The exploitation of frequencies above 30 GHz opens opportunities for high-speed wireless communication. Due to the limited maximum available power at such frequencies, techniques such as beam-forming and beam-steering have been widely used to both focus the transmitted beam and enable point-to-multipoint communication. The development of 60 GHz and 240 GHz data communication systems find place in this context.

5.1. *60 GHz beamforming circuits*

A beamforming 60 GHz system for communication purposes is being developed[56]. It consists of two different chips. The first one is a direct-conversion IQ transceiver with integrated integer-N PLL, capable of deliver/receive IQ signals within the four 60 GHz standard channels. The output of such a transceiver is thought to be connected to a bi-directional 8-channels beamforming chip, which will perform the gain and phase adjustment needed for beam steering. Its 8 inputs/outputs are intended to be connected to an 8 elements antenna array. The chip photo of the beamforming chip is shown in Fig. 11. An explicative system overview and the measured output polar plot are reported in Fig. 12.

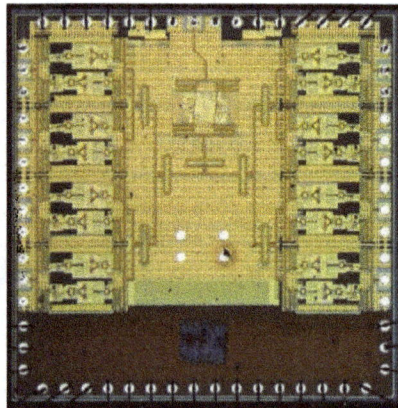

Fig. 11. Chip photo of the beam-forming chip.

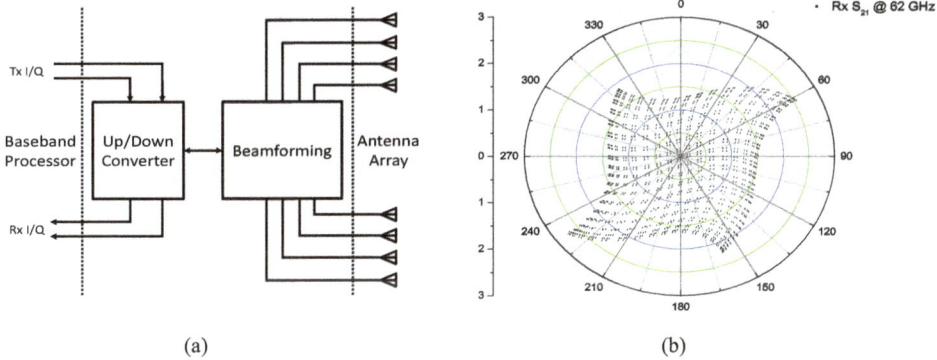

Fig. 12. System overview (a) and output polar plot (b) of the 60 GHz beamforming system.

5.2. *240 GHz short-range communication circuits*

In this section we report a 4-channel phased array system at 240 GHz using vector modulators and Butler matrix.

4-channel Vector modulator phased array

A 4-channel phased array system has been developed[57]. The input signal is divided to the four channels using various power dividers, both passives and actives. Then, each signal is fed into a vector modulator which allows a gain control of 18 dB and a phase control of 360^0. A chip photo of the system is reported in Fig. 13, while its block diagram and the measured output polar plot are shown in Fig. 14.

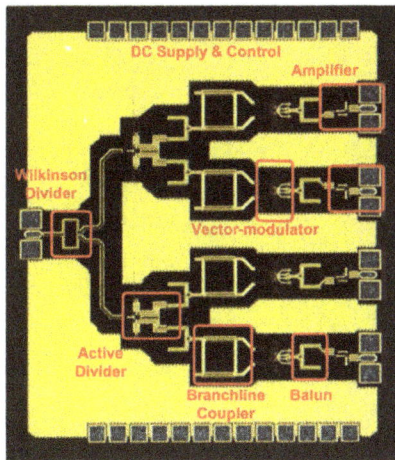

Fig. 13. Chip photo of the 4-channel Vector modulator phased array

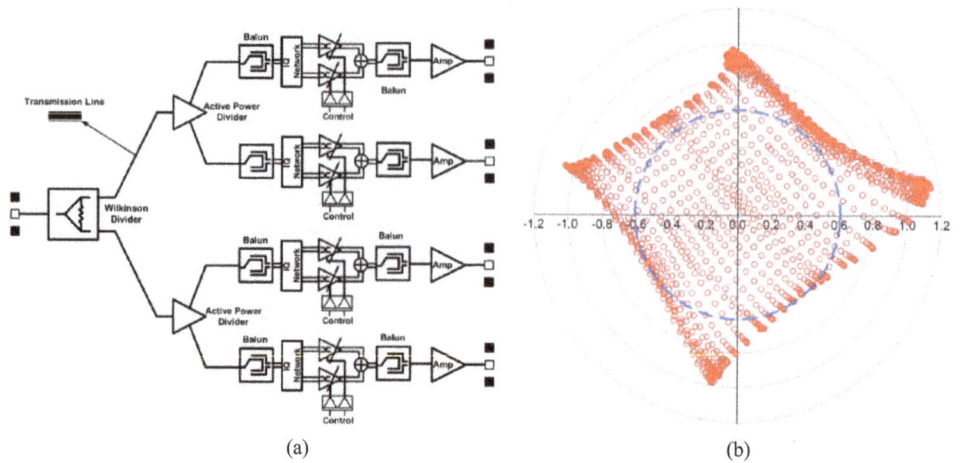

(a)   (b)

Fig. 14. System block diagram (a) and output polar plot (b) of the 240 GHz 4-channel vector modulator phased array.

4-Element Butler matrix

A 4-element Butler matrix beam switching chip has been developed as well[58]. The outputs of four LNAs are fed into a Butler matrix core with a single-pole-four-throw switch used to select between the different beam directions. A maximum phase error of 15^0 from the ideal phase states and less than 4 dB average amplitude variations have been obtained. The system chip photo is shown in Fig. 15, while in Fig. 16 can be seen its block diagram together with its measured array factor.

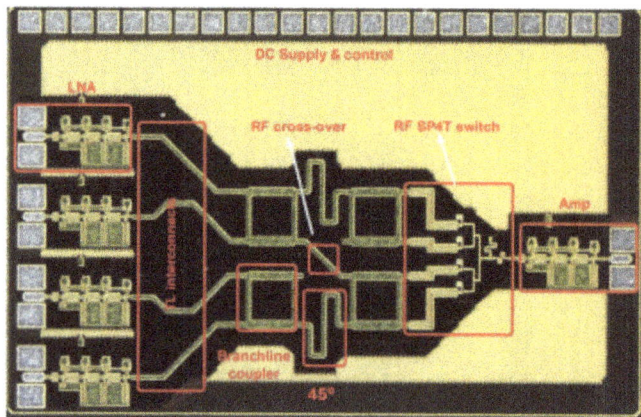

Fig. 15. Chip photo of the 4-element Butler matrix.

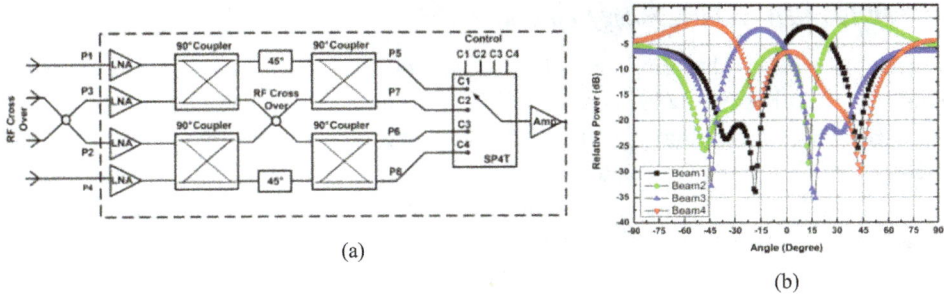

(a)

(b)

Fig. 16. System block diagram (a) and array factor (b) of the 240 GHz 4-element Butler matrix.

6. Millimeter-wave Radar Circuits

6.1. *120 GHz radar system-on-chip*

A highly-integrated 122 GHz system-on-chip radar sensor has been designed in IHP SiGe BiCMOS technology[59]. The block diagram as well as the micrograph of the implemented chip is shown in Fig. 17. The chip includes a radar transceiver and two on-chip antennas utilizing a novel antenna design approach that allows the use of the localized backside etching technique without compromising the mechanical stability of the chip. The implemented double folded dipole antenna achieves an antenna gain of 6 dBi with a radiation efficiency of 54%. The transceiver is equipped with a 61 GHz VCO that is complemented with a frequency doubler to generate the transmit signal. The receive path includes an LNA, a 90 degree coupler, two passive subharmonic mixers and variable gain amplifiers. Radar measurements with static as well as moving targets were done to show the applicability of the developed system.

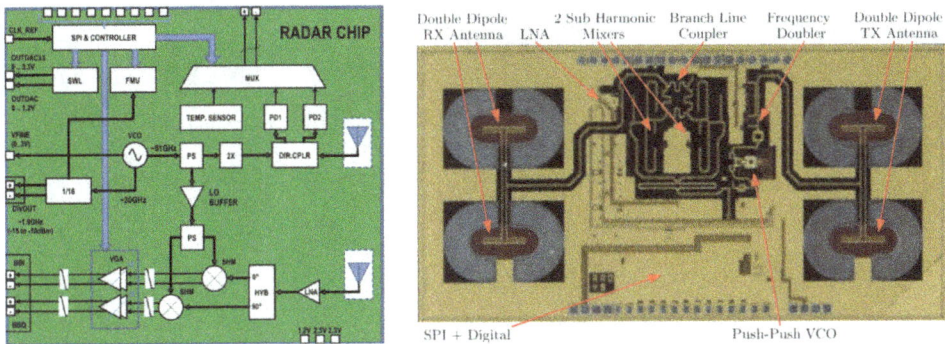

Fig. 17. Simplified block diagram and the micrograph of the 122 GHz SoC radar sensor consisting of 1-channel radar transceiver and two on-chip antennas.

6.2. 60/120 GHz scalable radar frontends

A scalable sensor platform consisting of several multi-purpose 61 and 122 GHz transceivers has been designed in a fully-differential architecture and implemented in IHP SiGe BiCMOS technology[60]. The block diagram of the platform as well as the micrographs of the implemented multi-band chips can be seen in Fig. 18. The former transceiver achieves a higher transmit output power as well as receive gain and is meant for applications requiring a high dynamic range, while the latter transceiver allows a higher modulation bandwidth and is meant for high resolution applications. The transceivers include a frequency multiplier that generates the 61 and 122 GHz carrier signals from a single external 30.5 GHz LO signal that is also fed to an output buffer in the transceivers. The proposed architecture enables the cascading of multiple transceivers and allows thus the implementation of a MIMO radar system with 2 different frequency bands. The transceivers are equipped with BPSK modulators as well as an I/Q receiver and can be utilized as a base to build a very flexible software-defined radar platform.

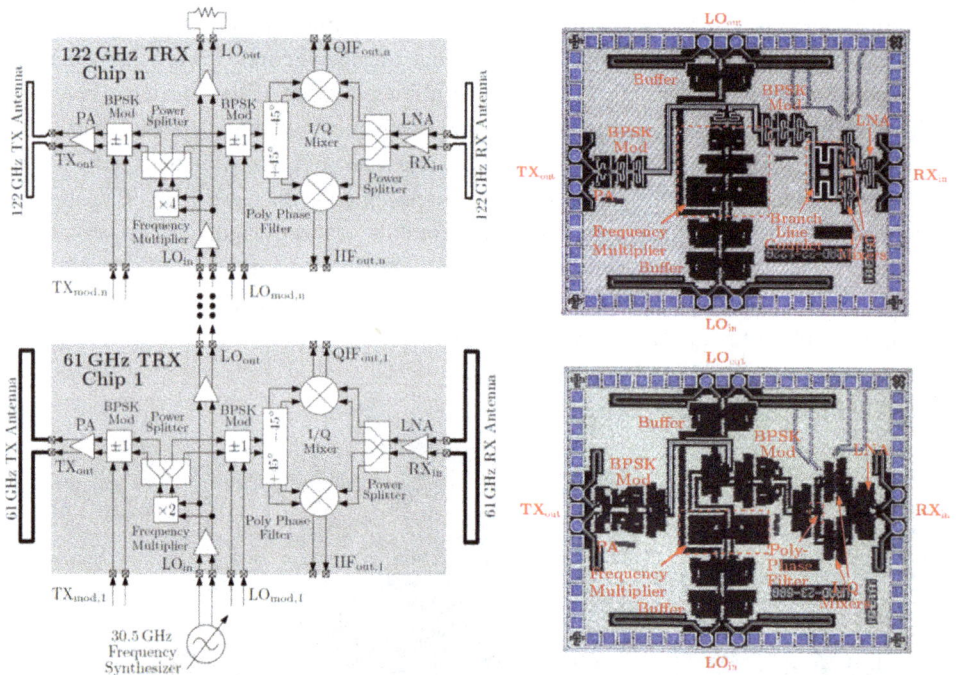

Fig. 18. Simplified block diagram and the micrographs of the proposed scalable radar sensor platform consisting of several multi-purpose 61 and 122 GHz transceivers that are cascaded to form a daisy chain.

7. THz Spectroscopy Circuits

The transmitter (TX) and receiver (RX) circuits used in a gas spectroscopy system are fabricated in IHP's 0.13 μm SiGe BiCMOS technology with f_T/f_{max} of 300 GHz/500 GHz.

7.1. *245 GHz TX/RX*

Our 245 GHz spectroscopic system for the frequency range 238–252 GHz includes a single TX and a subharmonic RX[61,62]. Further, a TX-array has been developed for the same frequency range. It delivers +7 dBm output power and 18 dBm effective isotropically radiated power (EIRP) at 245 GHz[63,64]. This array includes four TXs for spatial power combining. Each TX consists of a two-stage power amplifier (PA), a frequency doubler, and an integrated antenna, see Fig. 1. The inputs of these TXs are connected to a Wilkinson power divider network, which is fed by a LO. The LO consists of a 120 GHz push-push voltage controlled oscillator (VCO) with an 1/64 frequency divider for the fundamental frequency, a 120 GHz differential two-stage PA as used also for the TX, and an external phase lock loop (PLL). The 1/64 frequency divider is coupled inductively to the push-push oscillator core at the fundamental frequency. The PA uses two differential cascade stages with a transformer-coupled interstage topology. Fig. 19 shows the photograph of the 245 GHz TX-array chip.

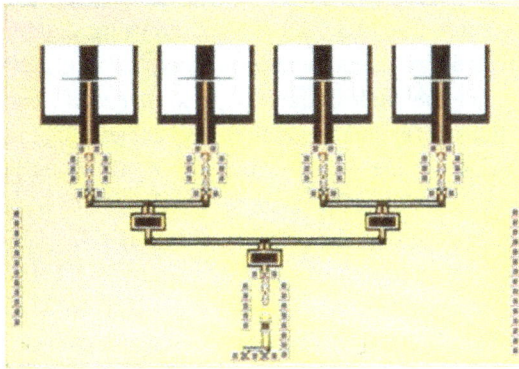

Fig. 19. Photograph of the 245 GHz TX- array chip.

7.2. *500 GHz TX/RX*

The 500 GHz TX-array includes frequency quadruplers instead of the frequency doublers of the 245 TX-array, see Fig. 2. The RX consists of a 122 GHz LO connected to a frequency doubler, and a transconductance subharmonic mixer connected to an integrated antenna. The die area of the fabricated TX-array is 3.3x2.6 mm^2 (Fig. 19), and the die area of the fabricated RX-chip is 2.3x0.9 mm^2. The output power of the TX-array is approximately -7 dBm at 500 GHz[65]. The tuning range of the TX-array is 480-500 GHz. We obtained for the RX without antenna a conversion gain (single ended) of -16 dB at 480 GHz RF signal and 1 GHz intermediate frequency (IF) signal[66]. The 3 dB IF-bandwidth is 3 GHz.

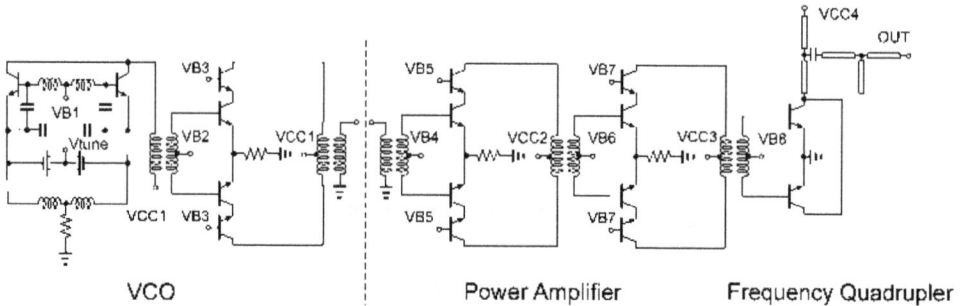

Fig. 20. Schematic of the VCO, and the two-stage PA coupled to the quadrupler.

8. Conclusion

In this work we reviewed the development and current performance status of SiGe-HBTs and BiCMOS technologies for mm-wave and THz high frequency applications. We discussed selected device modifications of the SiGe-HBT to achieve high frequency performances of a complex BiCMOS technology towards the 0.5 THz regime with respect to different technology generations of high-speed SiGe-BiCMOS processes at IHP. Moreover we depict the high-frequency performance difference with respect to maximum achievable transit (f_T) and oscillation frequencies (f_{max}) in comparison to RF-CMOS technologies and show the required increase of additional process effort for the HBT-module integration. Finally different high speed circuits which are based on the described technologies were presented like broadband ICs for optical communication, wireless communication circuits, mm-wave radar circuits and THz circuits operating at 500 GHz for spectroscopic applications. All reviewed results show the potential of SiGe-BiCMOS technologies for a broad range of high-speed applications.

Acknowledgments

The authors would like to thank the IHP cleanroom staff for wafer preparation, Gerhard Fischer for measurements, B. Heinemann and H. Rücker for helpful discussions and IHPs "Technology Department" and "Circuit Department" for excellent support and contributions to this work.

References

1. S. S. Iyer, G. L. Patton, S. L. Delage, S. Tiwari, and J. M. C. Stork, "Silicon-germanium base heterojunction bipolar transistors by molecular beam epitaxy", in *Tech. Dig. Int. Electron Device Meeting*, 1987, pp. 874-876.
2. Dan-Xia Xu, Guang-Di Shen, M. Willander, Wei-Xin Ni and G.V. Hansson, "n-Si/p-Si(1-x) Ge(x)/n-Si double-heterojunction bipolar transistors", *Appl. Phys. Lett*, 52, pp. 2239-2241, June 1988.
3. G. L. Patton, S. S. Iyer, S. L. Delage, S. Tiwari, and J. M. C. Stork, "Silicon–germanium-base heterojunction bipolar transistors by molecular beam epitaxy", IEEE Electron Device Lett., vol. 9, pp. 165-167, Apr. 1988.

4. G. L. Patton, J. H. Comfort, B. S. Meyerson, E. F. Crabbe, G. J. Scilla, E. de Fresart, J. M. C. Stork, J. Y.-C. Sun, D. L. Harame, and J. Burghartz, "75 GHz f_T SiGe base heterojunction bipolar transistors", IEEE Electron Device Lett., vol. 11, pp. 171-173, Apr. 1990.

5. D. L. Harame, E. F. Crabbe, J. D. Cressler, J. H. Comfort, J. Y.-C. Sun, S. R. Stiffler, E. Kobeda, J. N. Burghartz, M. M. Gilbert, J. Malinowski, and A. J. Dally, "A high-performance epitaxial SiGe base ECL BiCMOS technology", in *Tech. Dig. Int. Electron Device Meeting*, 1992, pp. 19-22.

6. D. L. Harame, J. M. C. Stork, B. S. Meyerson, K. Y.-J. Hsu, J. Cotte, K. A. Jenkins, J. D. Cressler, P. Restle, E. F. Crabbe, S. Subbanna, T. E. Tice, B. W. Scharf, and J. A. Yasaitis, "Optimization of SiGe HBT technology for high speed analog and mixed-signal applications", in Tech. Dig. Int. Electron Device Meeting, 1993, pp. 71-74.

7. E. Kasper, A. Gruhle, and H. Kibbel, "High speed SiGe-HBT with very low base sheet resistivity", in Tech. Dig. Int. Electron Device Meeting, 1993, pp. 79-81.

8. A. Schuppen, A. Gruhle, U. Erben, H. Kibbel, and U. Konig, "Multiemitter finger SiGe HBT's with fmax up to 120 GHz", in Tech. Dig. Int. Electron Device Meeting, 1994, pp. 377-380.

9. D. L. Harame, K. Schonenberg, M. Gilbert, D. Nguyen-Ngoc, J. Malinowski, S.-J. Jeng, B. S. Meyerson, J. D. Cressler, R. Groves, G. Berg, K. Tallman, K. Stein, G. Hueckel, C. Kermarrec, T. Tice, G. Fitzgibbons, K. Walter, D. Colavito, T. Houghton, N. Greco, T. Kebede, B. Cunningham, S. Subbanna, J. H. Comfort, and E. F. Crabbe, "A 200 mm SiGe-HBT technology for wireless and mixed signal applications", in Tech. Dig. Int. Electron Device Meeting, 1994.

10. A. Monroy, M. Laurens, M. Marty, D. Dutartre, D. Gloria, J. L. Carbonero, et al., "BiCMOS6G: A high performance 0.35 µm SiGe BiCMOS technology for wireless applications", IEEE BCTM Tech. Dig., pp. 121-124, 199

11. E. Ehwald, D. Knoll, B. Heinemann, K. Chang, J. Kirchgessner, R. Mauntel, et al., "Modular integration of high-performance SiGe:C HBT's in a deep submicron epi-free CMOS process", IEDM Tech. Dig., pp. 561-564, 1999.

12. G. Freeman, D. Ahlgren, D. R. Greenberg, R. Groves, F. Huang, G. Hugo, et al., "A 0.18 µm 90 GHz f_T SiGe HBT BiCMOS ASIC-compatible copper interconnect technology for RF and microwave applications", IEDM Tech. Dig., pp. 569-572, 1999.

13. S. Decoutere, F. Vleugels, R. Kuhn, R. Loo, M. Caymax, S. Jenei, et al., "A 0.35 µm SiGe BiCMOS process featuring a 80 GHz f_{max} HBT and integrated high-Q RF passive components", IEEE BCTM Tech. Dig., pp. 106-109, 2000.

14. K. Washio, E. Ohue, H. Shimamoto, K. Oda, R. Hayami, Y. Kiyota, et al., "A 0.2-µm 180-GHz-f_{max} 6.7-ps-ECL SOI/HRS self-aligned SEG SiGe HBT/CMOS technology for microwave and high-speed digital applications", IEDM Tech. Dig., pp. 741-744, 2000.

15. R. Lodge, "Advantages of SiGe for GSM RF front ends", Electron. Eng., vol. 71, no. 865, pp. 18-19, Feb. 1999 (UK).

16. R. Gotsfried, F. Beisswanger, S. Gerlach, A. Schuppen, H. Dietrich, U. Seiler, K. H. Bach, and J. Albers, "RFIC's for mobile communication systems using SiGe bipolar technology", *IEEE Trans. Microwave Theory, Tech.*, pt. 2, vol. 46, pp. 661-668, May 1998.

17. S. Subbanna, D. Ahlgren, D. Harame and B. Meyerson, "How SiGe evolved into a manufacturable semiconductor production process", IEEE ISSCC Dig. Tech. Papers, pp. 66-67, 1999.

18. Xi Li, T. Brogan, M. Esposito, B. Myers, et al., "A comparison of CMOS and SiGe LNA's and mixers for wireless LAN application", Conference on Custom Integrated Circuits, 2001, IEEE.

19. K. Washio "SiGe HBT and BiCMOS technologies for optical transmission and wireless communication systems" IEEE Transactions on Electron Devices, Volume: 50, Issue: 3, 2003.

20. T. Masuda, K. Ohhata, E. Ohue, K. Oda, M. Tanabe, H. Shimamoto, et al., "40 Gb/s analog IC chipset for optical receiver using SiGe HBTs", IEEE ISSCC Dig. Tech. Papers, pp. 314-315, 1998.

21. B. Heinemann; H. Rucker; R. Barth; J. Bauer; D. Bolze; E. Bugiel; J. Drews; K.-E. Ehwald; T. Grabolla; U. Haak; W. Hoppner; D. Knoll; D. Kruger; B. Kuck; R. Kurps; M. Marschmeyer; H. H. Richter; P. Schley; D. Schmidt; R. Scholz; B. Tillack; W. Winkler; D. Wolnsky; H.-E. Wulf; Y. Yamamoto; P. Zaumseil, "Novel collector design for high-speed SiGe:C HBTs", Internation Electron Devices Meeting IEDM, pp. 775-778, 2002.

22. T. Hashimoto, Y. Nonaka, T. Tominari, H. Fujiwara, K. Tokunaga, M. Arai, S. Wada, T. Udo, M.Seto, M. Miura, H. Shimamoto, K. Washio, H. Tomioka "Direction to improve SiGe BiCMOS technology featuring 200-GHz SiGe HBT and 80-nm gate CMOS" Electron Devices Meeting, 2003. IEDM '03 Technical Digest. IEEE International.

23. B. Heinemann et al., "SiGe HBT technology with fT/fmax of 300GHz/500GHz and 2.0ps gate delay", IEEE IEDM Techn. Dig., pp. 688-691, 2010.

24. H. Rücker et al., "Half-Terahertz SiGe BiCMOS Technology" Proc.Silicon Monolithic Integrated Circuits on RF Systems (SiRF 2012), pp. 133-136 (2012).

25. J.J. Pekarik et al., "A 90nm SiGe BiCMOS Technology for mm- wave and high-performance analog applications", Bipolar/BiCMOS Circuits and Technology Meeting (BCTM), 2014 IEEE.

26. P. Chevalier et al., "A 55 nm triple gate oxide 9 metal layers SiGeBiCMOS technology featuring 320 GHz fT / 370 GHz fMAX HBT and high-Q millimeter-wave passives" Proc. IEDM 2014, pp. 3.9.1-3.9.3.

27. B. Heinemann, H. Rücker, R. Barth, F. Bärwolf, J. Drews, G. Fischer, A. Fox, O. Fursenko, T. Grabolla, F. Herzel, J. Katzer, J. Korn, A. Krüger, P. Kulse, T. Lenke, M. Lisker, S. Marschmeyer, A. Scheit, D. Schmidt, J. Schmidt, A. Schubert, A. Trusch, C. Wipf, D. Wolansky, "SiGe HBT with fT/fmax of 505 GHz/720 GHz", accepted for IEDM 2016.

28. D. Harame et al., "mmWave RFCMOS Technology" WSG7-mmWave RFCMOS Technology and Applications (IMS (2015).

29. A. Mai et al., "SiGe-BiCMOS based technology platforms for mm-wave and radar applications", 21st International Conference on Microwave, Radar and Wireless Communications (MIKON), 2016.

30. B. Heinemann et al., "High-Performance BiCMOS Technologies withoutEpitaxially-Buried Subcollectors and Deep Trenches", Semiconductor Science and Technology vol. 22(1), pp. 153-157, (2007).

31. H. Rücker et al., "A 0.13 m SiGe BiCMOS Technology Featuring fT/fmax of 240/330 GHz and Gate Delays below 3ps" IEEE Journal of Solid State Circuits vol. 45(9), pp. 1678 (2010).

32. H. Rücker et al., "SiGe BiCMOS Technologies for Applications above 100 GHz", Proc. 2012 IEEE Compound Semiconductor Integrated Circuit Symposium (CSICS 2012), (2012) pp. 1-4.

33. J. S. Rieh, B. Jagannathan, H. Chen, K. T. Schonenberg, D. Angell, A. Chinthakindi, J. Florkey, F. Golan, D. Greenberg, S. J. Jeng, M. Khater, F. Pagette, C. Schnabel, P. Smith, A. Stricker, K. Vaed, R. Volant, D. Ahlgren, G. Freeman, K. Stein et al., "SiGe HBTs with cut-off frequency of 350 GHz", Electron Devices Meeting, 2002. IEDM '02 Technical Digest. IEEE International.

34. P. Chevalier et al., "High-speed SiGe BiCMOS technologies : 120-nm status and end-of Road challenges", Silicon Monolithic Integrated Circuits in RF Systems SiRF, 2007.

35. Wibo D. van Noort et al., "BiCMOS technology improvements for microwave applications", in Proc. Bipolar/BiCMOS Circuits and Technology Meeting, 2008.

36. B. Geynet, P. Chevalier, B. Vandelle, F. Brossard, N. Zerounian, M. Buczko, D. Gloria, F. Aniel, G. Dambrine, F. Danneville, D. Dutartre, and A. Chantre, "SiGe HBTs featuring > 400 GHz at room temperature", in Proc. Bipolar/BiCMOS Circuits and Technology Meeting, pp. 121-124, 2008.

37. A. Fox et al., "SiGe HBT Module with 2.5 ps Gate delay", IEEE IEDM Techn. Dig., 2008.

38. M. Khater et al., "SiGe HBT technology with f_T/f_{max}=350/300GHz and gate delays below 3.3ps", IEEE IEDM Techn. Dig., pp. 247-250, 2004.

39. E. Preisler et al., "A millimeter wave cable SiGe BiCMOS with 270 GHz f_{max} HBTS designed for high volume manufacturing", IEEE BCTM Proc., pp. 74-77, 2011.

40. P. Chevalier et al., "A conventional double-polysilicon FSA-SEG Si/SiGe:C HBT reaching 400 GHz f_{max}" IEEE BCTM Proc., pp. 70-73, 2009.

41. S. Van Huylenbroeck et al., "Pedestal collector optimization for high-speed SiGe:C HBTs", IEEE BCTM Proc. pp 66-69, 2011.

42. P. Chevalier et al., "Towards THz SiGe HBTs", IEEE BCTM Proc., BiCMOS Circuits and Technology Meeting (BCTM), 2011.

43. A. Fox et al., "SiGe:C Architecture with Epitaxial External base", IEEE BCTM Proc., pp. 70-73, 2011.

44. P. Candra et al., "130nm SiGe BiCMOS technology for mm-Wave applications featuring HBT with f_T/f_{max} of 260/320 GHz", Radio Frequency Integrated Circuits Symposium (RFIC), 2013 IEEE.

45. R. Lachner, "Towards 0.7 Terahertz Silicon Germanium Heterojunction Bipolar Technology – The DOTSEVEN Project", ECS Trans. 2014 volume 64, issue 6, 21-37.

46. Qizhi Z. Liua, James W. Adkissona, Vibhor Jain, Renata A. Camillo-Castillo, Marwan H. Khater, Peter B. Gray, John Jack Pekarik, Bjorn Zetterlund, Adam W. Divergilio, Michael L. Kerbaugh and D. L. Harame, "SiGe HBTs in 90nm BiCMOS Technology Demonstrating fT/fMAX 285GHz/475GHz through Simultaneous Reduction of Base Resistance and Extrinsic Collector Capacitance", ECS Trans. 2014 volume 64, issue 6, 285-294.

47. J. Böck et al., "SiGe HBT and BiCMOS Process Integration Optimization within the DOTSEVEN Project" Proc. IEEE Bipolar/BiCMOS Circuits and Technology Meeting (BCTM 2015), pp. 121-124 (2015).

48. J. Korn "Experimental and Theoretical Study of fT for SiGe HBTswith ascaled vertical doping profile", Bipolar/BiCMOS Circuits and Technology Meeting - BCTM, 2015 IEEE.

49. I. García López, A. Aimone, S. Alreesh, P. Rito, T. Brast, V. Höhns, G. Fiol, M. Gruner, J. K. Fischer, J. Honecker, A. G. Steffan, D. Kissinger, A. C. Ulusoy and M. Schell, "DAC-free Ultra-Low-Power Dual-Polarization 64-QAM Transmission at 32 GBd with Hybrid InP IQ SEMZM and BiCMOS Drivers Module", in *IEEE Journal of Lightwave Technology*, accepted.

50. I. García López, P. Rito, L. Zimmermann, D. Kissinger and A.C. Ulusoy, "A 40 Gbaud SiGe:C BiCMOS Driver for InP Segmented MZMs with Integrated DAC Functionality for PAM-16 Generation", in *2016 IEEE MTT-S International Microwave Symposium (IMS)*, May 2016, pp. 1-4.

51. P. Rito, I. García López, D. Petousi, L. Zimmermann, M. Kroh, S. Lischke, D. Knoll, D. Micusik, A. Awny, A. C. Ulusoy, and D. Kissinger, "A Monolithically Integrated Segmented Linear Driver and Modulator in EPIC 0.25 μm SiGe:C BiCMOS Platform", in *IEEE Transactions on Microwave Theory and Techniques, accepted.*

52. P. Rito, I. García López, D. Petousi, L. Zimmermann, M. Kroh, S. Lischke, D. Knoll, D. Kissinger, and A. C. Ulusoy, "A monolithically integrated segmented driver and modulator in 0.25 μm SiGe:C BiCMOS with 13 dB extinction ratio at 28 Gb/s", in *2016 IEEE MTT-S International Microwave Symposium (IMS)*, May 2016, pp. 1-4.

53. A. Awny, R. Nagulapalli, D. Micusik, J. Hoffmann, G. Fischer, D. Kissinger, and A. C. Ulusoy, "A Dual 64Gbaud 10kΩ 5% THD Linear Differential Transimpedance Amplifier with Automatic Gain Control in 0.13μm BiCMOS Technology for Optical Fiber Coherent Receivers", in *2016 IEEE International Solid-State Circuits Conference (ISSCC)*, San Francisco, CA, 2016, pp. 406-407.

54. A. Awny, R. Nagulapalli, G. Winzer, M. Kroh, D. Micusik, S. Lischke, D. Knoll, G. Fischer, D. Kissinger, A. C. Ulusoy and L. Zimmermann, "A 40 Gb/s Monolithically Integrated Linear Photonic Receiver in a 0.25 μm BiCMOS SiGe:C Technology", in *IEEE Microwave and Wireless Components Letters*, vol. 25, no. 7, pp. 469-471, July 2015.

55. M. Kroh, A. Awny, G. Winzer, R. Nagulapalli, S. Lischke, D. Knoll, A. Pęczek, D. Micusik, A. C. Ulusoy, D. Kissinger, L. Zimmermann and K. Petermann, "Monolithic Photonic-Electronic Linear Detection Receiver for 56Gbps OOK", in *2016 European Conference on Optical Communications (ECOC)*.

56. A. Malignaggi, M. Ko, A. C. Ulusoy, M. Petri, J. Gutierrez, E. Grass, and D. Kissinger, *Modular 60 GHz Beamforming Transceiver in 130-nm BiCMOS for Scalable 5G Backhaul Solutions*. Workshop: European Microwave Week, Oct. 2016, London, United Kingdom.

57. M. Elkhouly, S. Glisic, C. Meliani, F. Ellinger and J. C. Scheytt, *220 - 250-GHz Phased-Array Circuits in 0.13-um SiGe BiCMOS Technology*. IEEE *Transactions on Microwave Theory and Techniques*, vol. 61, no. 8, pp. 3115-3127, Aug. 2013.

58. M. Elkhouly, Y. Mao, C. Meliani, J. C. Scheytt and F. Ellinger, *A G-Band Four-Element Butler Matrix in 0.13 μm SiGe BiCMOS Technology* (*IEEE Journal of Solid-State Circuits*, vol. 49, no. 9, pp. 1916-1926, Sep. 2014.

59. H. J. Ng, J. Wessel, D. Genschow, R. Wang, Y. Sun, and D. Kissinger, "Miniaturized 122GHz system-on-chip radar sensor with on-chip antennas utilizing a novel antenna design approach", in *IEEE MTT-S Int. Microw. Symp. Dig.*, San Francisco, CA, May 2016.

60. H. J. Ng, M. Kucharski, and D. Kissinger, "Scalable sensor platform with multi-purpose fully-differential 61 and 122GHz transceivers for various MIMO radar applications", in *Proc. IEEE Bipolar/BiCMOS Circuits Technol. Meeting*, New Brunswick, NJ, Oct. 2016.

61. K. Schmalz, Y. Mao, J. Borngräber, P. Neumaier, H.-W. Hübers, "Tunable 245 GHz transmitter and receiver in SiGe technology for gas spectroscopy", Electronics Letters, vol. 50, no. 12, pp. 881-882, June 2014.

62. K. Schmalz, R. Wang, W. Debski, H. Gulan, J. Borngräber, P. Neumaier, H. W. Hübers, "245 GHz SiGe Sensor System for Gas Spectroscopy", Int. J. of Microwave and Wireless Technologies, vol.7 (3/4), pp. 271-278, June 2015.

63. K. Schmalz, J. Borngräber, W. Debski, M. Elkhouly, R. Wang, P. Neumaier, H.-W. Hübers, "245 GHz SiGe Transmitter Array for Gas Spectroscopy", in IEEE Proc. Compound Semiconductor Integrated Circuit Symposium (CSICS), San Diego, CA, USA, Oct. 2014, pp. 1-4.

64. K. Schmalz, J. Borngräber, W. Debski, M. Elkhouly, R. Wang, P. Neumaier, H.-W. Hübers, "245 GHz SiGe Transmitter Array for Gas Spectroscopy", IEEE Trans. on Terahertz Science and Technology, vol. 6, no. 2, pp. 318-327, March 2016.

65. K. Schmalz, J. Borngräber, W. Debski, P. Neumaier, R. Wang, H. W. Hübers, "Tunable 500 GHz transmitter-array in SiGe technology for gas spectroscopy", Electronics Letters, vol. 51, no. 3, pp. 257-259, Feb. 2015.

66. K. Schmalz, J. Borngräber, W. Debski, P. Neumaier, R. Wang, D. Kissinger, and H.-W. Hübers, "Tunable 500 GHz sensor system in SiGe technology for gas spectroscopy", Electronics Letters, vol. 51, no. 17, 2015, pp. 1345-1347, Aug. 2015.

Optimization of Selective Growth of SiGe for Source/Drain in 14nm and Beyond Nodes FinFETs

Henry H. Radamson[*,†,‡,§], Jun Luo[*,†], Changliang Qin[*,†], Huaxiang Yin[*,†], Huilong Zhu[*,†], Chao Zhao[*,†] and Guilei Wang[*,†,**]

[*]Key laboratory of Microelectronics Devices & Integrated Technology, Institute of Microelectronics, Chinese Academy of Sciences, Beijing, 100029, P. R. China
[†]University of Chinese Academy of Sciences, Beijing 100049, P. R. China
[‡]KTH Royal Institute of Technology, Brinellv. 8, 10044 Stockholm, Sweden
[§]rad@ime.ac.cn
[**]wangguilei@ime.ac.cn

In this work, optimization of selective epitaxy growth (SEG) of SiGe layers on source/drain (S/D) areas in 14nm node FinFETs with high-k & metal gate has been presented. The Ge content in epi-layers was in range of 30%-40% with boron concentration of $1-3 \times 10^{20}$ cm^{-3}. The strain distribution in the transistor structure due to SiGe as stressor material in S/D was simulated and these results were used as feedback to design the layer profile. The epitaxy parameters were optimized to improve the layer quality and strain amount of SiGe layers. The in-situ cleaning of Si fins was crucial to grow high quality layers and a series of experiments were performed in range of 760-825 °C. The results demonstrated that the thermal budget has to be within 780-800 °C in order to remove the native oxide but also to avoid any harm to the shape of Si fins. The Ge content in SiGe layers was directly determined from the misfit parameters obtained from reciprocal space mappings using synchrotron radiation. Atomic layer deposition (ALD) technique was used to deposit HfO$_2$ as high-k dielectric and B-doped W layer as metal gate to fill the gate trench. This type of ALD metal gate has decent growth rate, low resistivity and excellent capability to fill the gate trench with high aspect-ratio. Finally, the electrical characteristics of fabricated FinFETs were demonstrated and discussed.

Keywords: SiGe; selective epitaxy growth; FinFET.

1. Introduction

For many years the evolution of CMOS technology was performed through down-scaling of the transistor physical dimensions in order to nearly double their density in the chip every other year. As these developments continued to 22nm node a revolutionary step in the transistor design occurred and the traditional planar transistors converted to 3D transistors e.g. FinFET. The new design improves the transistor performance by exerting an electrostatic control to the transistor channel and decreases the short-channel effects [1-7]. For such structures, SiGe layers were grown selectively on source/drain

---

[*,**]Corresponding authors.

regions as stressor material to create uniaxial strain in the channel region and to enhance the carrier mobility [1-7].

The quality of epi-layer is sensitive to both growth parameters and the presence of residuals of native oxide or impurities on the surface of Si fins. Therefore, both ex-situ and in-situ cleaning are important steps to provide Si fins which are free of undesired atoms.

In FinFETs, the shape, dimensions (height and width) and doping level of the original Si fin have importance in carrier transport [2, 8-9]. The shape of fin plays the role how stress from SiGe is exerted from sidewalls to the central part of body. To process the Si fin, an anisotropic is applied and depending on how long the over-etch is performed the fin shape can be designed to be trapezoid or triangular form.

This work presents the optimization of selective epitaxy growth of highly strained SiGe layers in 14nm node FinFETs with integrated high-k & metal gate. Moreover, the strain in transistor structure was simulated and transistor performance was characterized and discussed.

2. Experimental Details

The FinFETs were processed on 200mm Si (100) wafers. Sidewall transfer lithography (STL) technique with an anisotropic dry etch was applied to create Si fins with 110 nm height and 15 nm width [10]. In this step, Si_3N_4/SiO_2 spacers were used as hard mask for plasma-etch of the Si. Later, these spacers were removed in hot solution of H_3PO_4 and diluted HF.

Since there is a huge risk that the fins could suffer from damages during the plasma etch an oxidation step was performed in a rapid thermal processing (RTP) chamber. This oxidation step can make the corners of Si fins round which is beneficial shape for strain engineering. Later, a 200nm-thick oxide layer was grown as shallow trench isolation (STI) oxide and the topography was planarized by chemical mechanical polishing (CMP). The STI oxide was recessed by applying a combination of dry etch and wet etch in a diluted HF solution. An amorphous Si was deposited and patterned by electron beam lithography (EBL) crossing over the fins. Later spacers were formed at both sides of the patterned Si finalizing the dummy gate for transistors. A standard cleaning procedure (SPM followed by APM with DHF at last) was applied to clean the samples prior to epitaxy. These wafers were placed into the load-locks of a reduced pressure chemical vapor deposition (RPCVD) reactor. An in-situ cleaning was performed by baking the samples at temperatures in range of 760-825 °C for 7 min. The SiGe layers were grown on the Si-fins at 650 °C with total pressure of 20 torr. In the epitaxy process, SiH_2Cl_2, 10% GeH_4 in H_2 and 1% B_2H_6 in H_2 were the precursors of Si, Ge and B, respectively. HCl was employed to remove the formed nuclides on oxide (or nitride) and to ensure the selectivity during the deposition. After the SiGe epitaxy, pre-metal dielectric (PMD) was deposited and the dummy gate was opened by CMP. Diluted TMAH (Tetramethylammonium Hydroxide) solution followed by a HF-dip was used to etch the dummy gate and to remove the interfacial oxide. In this step, the real gate oxide was a 2.4nm-thick HfO_2 which was deposited by atomic layer deposition technique. Later, a metal stack of TiN/Ti/TiN was deposited as work function metals for transistors by using ALD, PVD and CVD techniques, respectively. The gate trench was filled by a 100nm-thick B-doped W film was deposited by using ALD technique. A summary of process flow, a schematic feature and cross-section TEM of FinFETs pMOS with deposited SiGe on source/drain regions are shown Fig. 1a to 1c.

a)

b)

c)

- Fin formation
- Dummy gate formation
- S/D SEG
- New spacers
- HKMG, All-last Module
- Contact formations

Spacer Gate SiGe SEG STI Si substrate

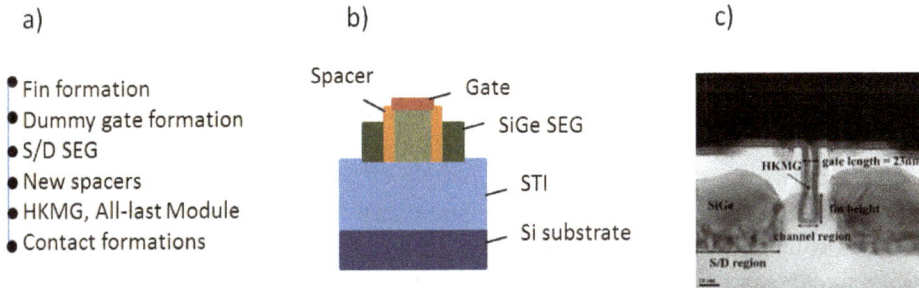

Fig. 1. (a) Process flow sequence, (b) a schematic view of Si fin/SiGe and (c) is a cross-sectional TEM image of a fully processed transistor.

The strain amount (or Ge content) in the processed FinFET structures was determined by high-resolution reciprocal lattice mapping (HRRLM) at (-115) reflection by using x-ray beam in a Synchrotron facility. The high intensity beam could provide the opportunity to measure the misfit parameters and therefore the strain relaxation for small SiGe crystals [11-12].

The boron concentration was calculated from HRRLMs by estimating the shift of B-doped SiGe peak compared with intrinsic one due to strain compensation.

The strain distribution in the FinFET structures was simulated by using Sentaurus TCAD of Synopsys. This simulation program applied drift-diffusion model where the TCAD software itself was initiated from Atomistic Kinetic Monte Carlo Diffusion Method.

3. Result and Discussions

The design of 3D transistors such as FinFETs differs significantly from 2D planar transistors. The most important design parameters for the FinFETs are the shape of Si-fin, doping level in the transistor body and the strain amount in the channel region. These parameters affect carrier mobility and carrier profile as well as the threshold voltage of transistors [9].

SiGe material is used as stressor materials in S/D regions in FinFET structures. Selective epitaxy growth is applied to raise the S/D inducing strain in the channel region. This type of epitaxy suffers from a problem so-called pattern dependency which yields to different SiGe profiles across the transistor arrays. This behavior occurs when the density and size of the transistor vary in a chip [13-19]. The reason behind the pattern dependency of SEG is non-uniform consumption of reactant gas molecules when the exposed Si area varies over the chip. In this work, the effect of this pattern dependency for Ge content was estimated to 10-15% whereas the layer thickness could change up to 200% from the center to edge of 200mm wafer.

So far, different methods have been proposed to decrease this non-uniformity of deposition but eliminating this behavior completely has never been demonstrated. Therefore, simulation of strain from different SiGe profiles is necessary to provide information about the variation of transistor performance due to the pattern dependency of growth.

The strain distribution in the FinFET structures was simulated by using input parameters: fins' dimensions 35nm × 10nm × 20nm (height × top width × bottom width), SiGe profile, spacer width and gate length.

Figure 2 shows the simulation results of strain distribution in the transistor structure. The strain is uniformly distributed in the channel region with strain amount of 0.4 and 0.8 GPa for 30% and 40% Ge content, respectively.

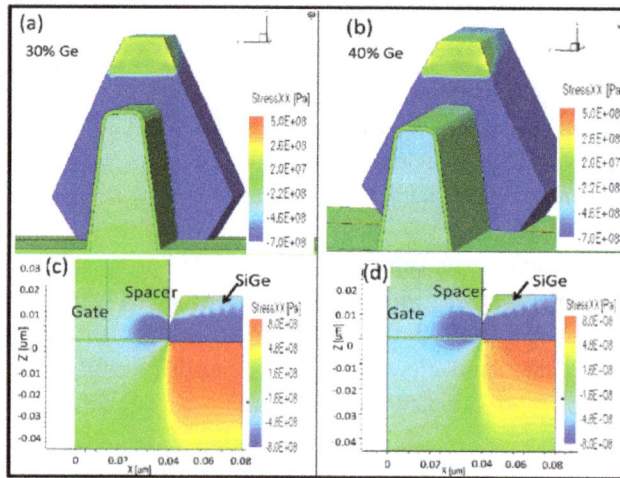

Fig. 2. Strain simulation results for the 14 nm node Si fin with $Si_{0.60}Ge_{0.30}$ (a, c) and $Si_{0.60}Ge_{0.40}$ (b, d) in source/drain regions.

This simulated stain distribution is for an ideal case and in order to induce such amount of strain by SiGe layers, the layer quality has to be high. In this case, the initial surface of Si fin has to be free of native oxide or any residual of impurities. Therefore, a focused work was spent on in-situ cleaning. Fig. 3a-d show the high-resolution transmission electron microscope (HRTEM) and scanning electron microscope (SEM) results from a series of experiments on pre-baking in temperature range of 760-825 °C. In Fig. 3a, the shape of Si fin is affected where their height is shrunk after pre-baking at

825 °C. Nevertheless the shape the Si fins is changed the SiGe layer could be grown problem free on the facet planes of Si fins. The change of fin shape could be described by exposure of small amount of HCl which usually exists in the deposition-line. HCl is Si etchant and the etch process is temperature dependent (kinetic mode in CVD). The other plausible reason behind the change of Si fin is thermal mismatch between Si and oxide. Since the Si-fin has a tiny size the exerted force could be huge enough to deform the shape.

Therefore, the thermal budget was decreased to 800 °C and later to 780 °C where the Si fins shape is less unaffected in Fig. 3b and 3c. Further decrease of prebaking temperature to 760 °C cannot remove the native oxide resulting in irregular shape of SiGe layers with the large surface roughness (see Fig. 3d).

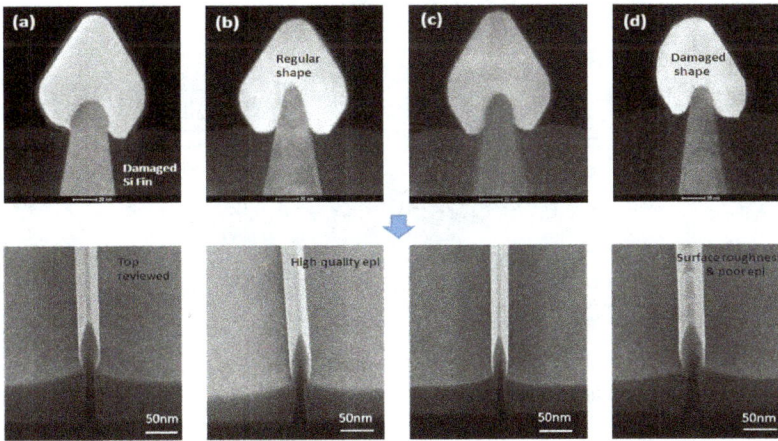

Fig. 3. Cross-sectional TEM images of Si Fin/SiGe samples together with their top view SEM images with different prebaking temperatures as following: (a) at 825 °C, (b) 800 °C, (c) 780 °C, and (d) 760 °C.

The other important for epitaxy was selectivity of the growth. A series of different HCl partial pressures was applied during the SiGe epitaxy and the selectivity, surface roughness and strain relaxation of the epi-layers were investigated. Figs. 4a-c illustrates the cross-sectional images from SiGe/Si fins grown with different HCl partial pressures. When the HCl amount is not sufficient (50 mtorr) and nucleation occur on the gate sidewalls (Fig. 4a) however when this value was increased to 60 mtorr a good selectivity could be achieved (Fig. 4b). The growth rate of this selectively grown SiGe was 50 nm/min which is slightly lower than 65 nm/min for the SiGe layer in Fig. 4a. Further increase of HCl partial pressure decreases the growth rate strongly and at last no epitaxy occurs on Si fins (Fig. 4c) for high HCl pressure (70 mtorr) as shown in Fig. 4c.

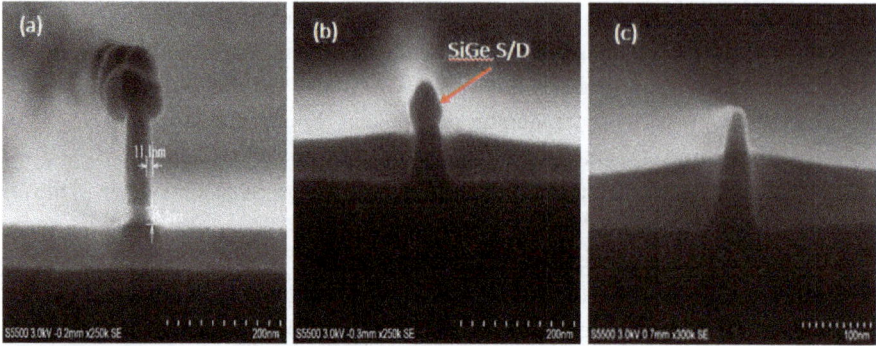

Fig. 4. Cross-sectional images of Si Fin/SiGe samples grown at 650 °C with DCS and GeH_4 partial pressures of 120 and 40 mTorr, respectively, when HCl partial pressures were (a) 50 mtorr, (b) 60 mtorr, and (c) 70 mtorr.

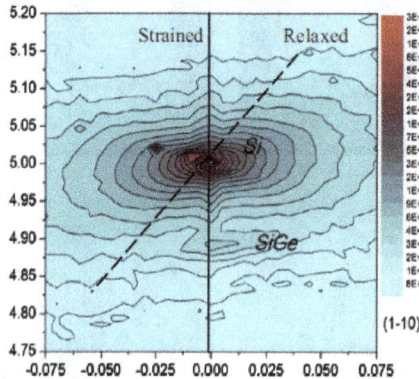

Fig. 5. HRRLM at (-115) reflection from an array of FinFETs located at the center of 8-inch wafer.

The other important issue after SiGe deposition was formation of HK&MG in the gate region. At first, HfO_2 was deposited as gate oxide [20] and later the trench was filled out by B-doped W layer. For both cases, ALD technique was used which demonstrates uniform high-k deposition as well as the gate metal could fill the gate trench excellently. [21, 22]. At this stage, the wafers were transferred to a synchrotron facility to measure strain (or Ge content) in the epi-layers. Fig. 5 shows a HRRLM around (-115) from 14nm FinFET arrays. The SiGe peak is aligned with Si along K// indicating minor strain relaxation. The mismatch parameters were obtained from this HRRLM and a parabolic relation was used to determine the lattice constant of SiGe layer and the Ge content [2, 11-12]. These calculations showed highly strained $Si_{0.60}Ge_{0.40}$ layers in FinFETs.

At final step, the electrical properties of the fully processed transistors were investigated as shown in Fig. 6a and 6b.

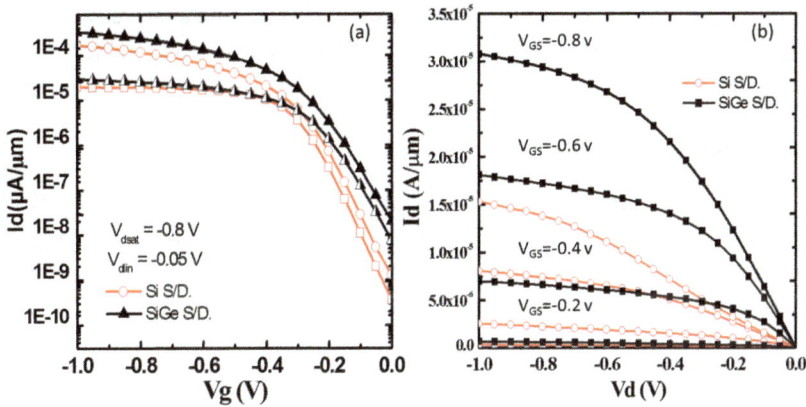

Fig. 6. Output characteristics curves of 14 nm FinFET devices with SiGe S/D compared to Si S/D (a) $I_d (V_g)$ and (b) $I_d (V_d)$.

Although, the parasitic resistances of S/D junctions were not optimized [22], the characteristics of transistor lied in the reasonable range. For example, a I_{on}-value of 292 µA/µm) at $V_g = V_d = -0.8$ V was measured where I_{off}, V_{sat}, drain-induced barrier lowering (DIBL) and subthreshold swing (SS) values are better compared to Si S/D (Fig. 6a). The FinFET design with SiGe S/D provides a controlled short channel effects (SCEs) as observed the threshold voltage (Vt), DIBL and SS were −0.24 V, 50 mV/V and 72 mV/dec, respectively. However a FinFET with Si source and drain had a performance of threshold voltage (V_t), DIBL and SS were -0.34 V, 50 mV/V and 65 mV/dec. The carrier mobility in the transistors with SiGe S/D was 88 cm²/V·s which is only slightly improvement compared to FinFETs without SiGe layers in S/D regions (~70 cm²/ V·s). A possible explanation for low mobility values could still be related to the unoptimized parasitic resistances of S/D junctions. Formation of silicides could solve these problems and the transistor could be improved [23]. In all cases, the FinFETs with SiGe in S/D have a better performance compared to Si ones due to two main reasons: the presence of strain in the channel region and lower the contact resistance due to SiGe layer.

4. Conclusions

The selective epitaxy of SiGe layers ($0.30 \leqslant x \leqslant 0.40$) in S/D regions was integrated for 14 nm node FinFETs with HK&MG. It was demonstrated that the thermal budget has to be in range of 780-800 °C range in order to avoid damages to the Si-fin shape and to grow SiGe layers with high epitaxial quality. HCl partial pressure was optimized during the epitaxy to obtain good selectivity with decent growth rate. The strain distribution due to SiGe in S/D was simulated by drift-diffusion model using TCAD software which is initiated from Atomistic Kinetic Monte Carlo Diffusion Method. The induced strain by SiGe layer was experimentally measured by HRRLMs using synchrotron beam focused on an array of FinFETs. The processed FinFETs demonstrated good characteristics in terms of I_{on}, I_{off}, V_{sat}, drain-induced barrier lowering (DIBL) and subthreshold swing (SS) values. In all cases, the FinFETs with SiGe in S/D demonstrated a better performance compared to Si ones.

Acknowledgments

This work was financially supported by "National S&T Major Project 02", the opening project of Microelectronics Devices & Bulk Si FinFET Integrated Technology, Institute of Microelectronics, Chinese Academy of Sciences (Project No. 2013ZX02303007-001) and "National Key Research and Development Program of China" (2016YFA0301701), which are hereby acknowledged.

References

[1] T. Chiarella, L. Witters, A. Mercha, C. Kerner, R. Dittrich, M. Rakowski, C. Ortolland, L.-A. Ragnarsson, B. Parvais, A. De Keersgieter, S. Kubicek, A. Redolfi, R. Rooyackers, C. Vrancken, S. Brus, A. Lauwers, P. Absil, S. Biesemans, and T. Hoffmann, in 2009 Proc. Eur. Solid State Device Res. Conf. pp. 85-88 (2009).
[2] H. Radamson and L. Thylen, Monolithic Nanoscale Photonics-electronics Integration in Silicon and Other Group IV Elements [M]. Elsevier Science & Technology (2014).
[3] G. L. Wang, M. Moeen, A. Abedin, M. Kolahdouz, J. Luo, C.L. Qin, H. L. Zhu, J. Yan, H. Z. Yin, J. F. Li, C. Zhao, and H. H. Radamson, J. Appl. Phys. vol. 114, p. 123511 (2013).
[4] C. Qin, G. Wang, M. Kolahdouz, J- Luo, H. Yin, P. Yang, J. Li, H. Zhu, Z. Chao, T. Ye, H. H. Radamson, Solid-State Electronics vol. 124, pp. 10-15 (2016).
[5] S. E. Thompson, R. S. Chau, T. Ghani, K. Mistry, S. Tyagi, and M. T. Bohr, IEEE Trans. Semicond. Manuf. vol. 18, p. 26 (2005).
[6] H. H. Radamson and M. Kolahdouz, Journal of Materials Science: Materials in Electronics vol. 26, pp. 4584-4603 (2014).
[7] T. Ghani, M. Armstrong, C. Auth, M. Bost, P. Charvat, G. Glass, T. Hoffmann, K. Johnson, C. Kenyon, J. Klaus, B. McIntyre, K. Mistry, A. Murthy, J. Sandford, M. Silberstein, S. Sivakumar, P. Smith, K. Zawadzki, S. Thompson, and M. Bohr, in IEEE Int. Electron Devices Meet. pp. 11.6.1-11.6.3 (2003).
[8] G. Pei, J. Kedzierski, P. Oldiges, M. Ieong, and E.C.-C. Kan, IEEE Trans. Electron Devices vol. 49, pp. 1411-1419 (2002).
[9] R. T. Buhler, J. A. Martino, P. G. D. Agopian, R. Giacomini, E. Simoen, and C. Claeys, in 2010 IEEE Int. SOI Conf. pp. 1-2 (2010).
[10] Hallstedt, P.-E. Hellstrom, H. H. Radamson, Thin Solid Films vol. 517, pp, 117-120 (2008).
[11] G. V. Hansson, H. H. Radamsson, and W.-X. Ni, Journal of Materials Science: Materials in Electronics vol. 6, pp. 292-297 (1995).
[12] H. H. Radamson and J. Hållstedt, Journal of Physics: Condensed Matter vol. 17 pp. 2315-2322 (2005).
[13] R. Loo and M. Caymax, Appl. Surf. Sci. vol. 224, pp. 24-30 (2004).
[14] J. Hartmann, L. Clavelier, C. Jahan, P. Holliger, G. Rolland, T. Billon, and C. Defranoux, J. Cryst. Growth vol. 264, pp. 36-47 (2004).
[15] M. Kolahdouz, L. Maresca, R. Ghandi, A. Khatibi, and H. H. Radamson, ECS pp. 581-593 (2010).
[16] J. Hållstedt, M. Kolahdouz, R. Ghandi, H. H. Radamson, and R. Wise, J. Appl. Phys. vol. 103 p. 054907 (2008).

[17] M. Kolahdouz, L. Maresca, M. Ostling, D. Riley, R. Wise, and H. H. Radamson, Solid State Electronics vol. 53, pp. 858-861 (2009).

[18] S. Bodnar, E. De Berranger, P. Bouillon, M. Mouis, T. Skotnicki, and J. Regolini, Journal of Vacuum Science & Technology B vol. 15, pp. 712-718 (1997).

[19] R. Loo, G. Wang, L. Souriau, J. Lin, S. Takeuchi, G. Brammertz, et al., ECS Transactions vol. 25, pp. 335-350 (2009).

[20] M. Johansson, M. Y. A. Yousif, P. Lundgren, S. Bengtsson, J. Sundqvist, A. Harsta, and H. H. Radamson, Semiconductor Science and Technology vol. 18, pp. 820-826 (2003).

[21] G. Wang, Q. Xu, T. Yang, J. Xiang, J. Xu, J. Gao, C. Li, J. Li, J. Yan, D. Chen, T. Y. C. Zhao, and J. Luo, ECS Journal of Solid State Science and Technology vol. 3, pp. 82-85 (2014).

[22] Q. Xu, J. Luo, G. Wang, T. Yang, J. Li, T. Ye, D. Chen, and Chao Zhao, Microelectronic Engineering vol. 137, pp. 43-46 (2015).

[23] O. Nur, M. Willander, L Hultman, H. Radamson, and G. V. Hansson, Journal of Applied Physics vol. 78, pp. 7063-7069 (1995).

Dynamic Conductivity and Two-Dimensional Plasmons in Lateral CNT Networks

Maxim Ryzhii

Department of Computer Science and Engineering, University of Aizu
Aizu-Wakamatsu 965-8580, Japan
m-ryzhii@ieee.org

Taiichi Otsuji

Research Institute of Electrical Communication, Tohoku University
Sendai 980-8577, Japan

Victor Ryzhii

Research Institute of Electrical Communication, Tohoku University
Sendai 980-8577, Japan and
Institute of Ultra High Frequency Semiconductor Electronics of RAS
Moscow 117105, Russia

Vladimir Mitin

Department of Electrical Engineering, University at Buffalo
Buffalo, NY 1460-1920, USA

Michael S. Shur

Departments of Electrical, Electronics, and Systems Engineering
Rensselaer Polytechnic Institute, Troy, NY 12180, USA

Georgy Fedorov and Vladimir Leiman

Department of General Physics, Moscow Institute of Physics and Technology
Dolgoprudny 141700, Russia

We study theoretically the carrier transport and the plasmonic phenomena in the gated structures with dense lateral carbon nanotube (CNT) networks (CNT "felt") placed between the highly-conducting slot line electrodes. The CNT networks under consideration consist of a mixture of semiconducting and metallic CNTs. We find the dispersion relations for the two-dimensional plasmons, associated with the collective self-consisted motion of electrons in the individual CNTs, propagating along the electrodes as functions of the net electron density (gate voltage), relative fraction of the semiconducting and metallic CNTs, and the spacing between the electrodes. In a wide range of parameters, the characteristic plasmonic frequencies can fall in the terahertz (THz) range. The structures with lateral CNT networks can used in different THz devices.

Keywords: carbon nanotube network; two-dimensional carrier system; terahertz radiation; plasmonic.

1. Introduction

Plasmonic properties of the heterostructures with two-dimensional (2D) electron or hole systems (2DES and 2DHS) can be used in different terahertz (THz) devices.[1,2] The structures with carbon nanotubes (CNTs) can also exhibit plasmonic response. In particular, the detection of THz radiation in lateral device structures based on CNTs was reported.[3–7] Different CNT-based detectors and mechanisms of their operation were reviewed in Ref. 8.

The plasmonic nature of the THz response in CNT systems was discussed in a number of publications (see, for example,[9–13]). The role of the plasmonic effects in aligned CNTs and disordered CNT networks) was experimentally confirmed in particularly in Refs. 14–17. Usually, the experimental results were interpreted in the framework of the concept of 1D plasmons propagating along the individual CNT and having cylindrical symmetry of the potential distribution(see, for example,[10]). However, as shown recently,[18] in sufficiently dense lateral CNT networks the carriers in the CNT can effectively interact with the carriers in many other CNTs via the self-consisted electric field. In this case, the plasmons in such CNT network have the 2D nature and characterized by the dispersion more specific for 2DESs and 2DHSs. Naturally, the anisotropy of the carrier in-plane motion and angular distribution of the CNTs can play a marked role in the dispersion characteristics of such 2D plasmons.

In this paper, we theoretically study the 2D plasma waves in the gated lateral dense CNT disordered networks (which can be called as the CNT "felts") consisting of a mixture of the single-wall semiconducting CNTs (s-CNTs) and metallic CNTs (m-CNTs) places between the highly-conducting strip-like electrodes, which constitute the slot line.

2. Device Model

The CNT network constitutes the CNT "felt" consisting of randomly oriented sufficiently long CNTs ($\mathcal{L} \gg L$, where $\mathcal{L}$ and L are the characteristic length of the CNTs and the spacing between the electrodes, respectively). The CNT network is attracted to the SiO_2 layer by the Van-der-Waals forces. Figure 1 shows schematically the device structure under consideration. The structure is doped with acceptors. The highly conducting substrate (doped Si), separated from the CNT network by a barrier layer (SiO_2), plays the role of the gate. The gate voltage V_g the population of both s- and m-CNTs with carriers (holes) and the average sheet charge density in the CNT network. The ratio of the densities (per unit length in the

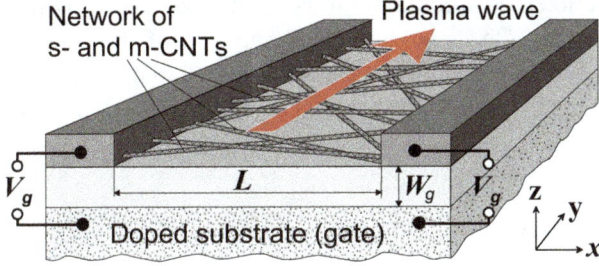

Fig. 1. Schematic view of the gated structure with a lateral CNT network.

in-plain direction parallel to the contacts) of the m-CNTs and s-CNTs, N_m and N_s respectively, can vary in a wide range starting from $N_m/N_s = 1/2$ to zero. The latter case can be realized, for example, by the scorching (burning) of the m-CNTs by strong electric pulses.

We assume that the carrier (hole) dispersion in the s-CNTs is given by a one-dimensional parabolic energy dependence on the momentum p with the effective mass m_s: $\varepsilon = \Delta_g/2 + p^2/2m_s$, whereas the dispersion of m-CNTs is by the quasi-relativistic dispersion relation which in the case of small energy gap can be presented as $\varepsilon \simeq v_W p$. Here Δ_g is the energy gap in the s-CNTs and $v_W \simeq 10^8$ cm/s is the characteristic velocity of the gapless spectrum.[19] We consider the sufficiently dense CNT networks with a large number $\mathcal{N}$ of the holes in the area with the radius equal to the Debye screening length l_D. In this situation, the lateral carrier movementum is associated with the effect of the self-consistent electric field. However, the average distance between the CNTs N_m^{-1}, N_s^{-1} is assumed to be much larger than the size of the localization region around individual CNTs, so that the wave function overlap is insignificant. This allows us to consider the carrier system in each CNT as one-dimensional. The pertinent estimates will be provided in the following.

The ac components of the electric field (external or self-consistent, associated with the ac variations of the hole density) result in the variation of hole distribution function $\delta f_\omega(p, \alpha)$ (where p is the hole momentum along the CNT and α is the angle between the direction perpendicular to the electrodes and the CNT axis) which can be found from the kinetic (Boltzmann) equation:

$$(i\omega + \nu_{s,m})\delta f_\omega^{s,m}(p, \alpha) = -e\left(\cos\alpha \frac{d\delta\varphi_\omega}{dx} + \sin\alpha \frac{d\delta\varphi_\omega}{dy}\right)\frac{df_0^{s,m}(p)}{dp}. \qquad (1)$$

Here ν is the phenomenological hole scattering frequency due to the hole scattering with disorder ($\nu = \nu_s$ and $\nu = \nu_m$ in the semiconducting and quasi-metallic CNTs, respectively), $f_0^{s,m}(p)$ are the equilibrium Fermi distribution function of the holes in the s-CNTs and m-CNTs, respectively, and $e = |e|$ is the elementary charge. The axis x is directed perpenticular to the strip-like electrodes and the axis y is directed along these electrodes in the CNT network plane. In sufficiently dense CNT networks, the intersections of the CNTs can occur. In the model under consideration, the CNT intersections are accounted for as additional scattering points for the holes propagating along the CNTs leading to an increase of the frequencies ν_m and ν_s. For simplicity, we disregard the polarization associated with the displacement of the holes perpenticular the CNTs, because it can be crucial only at very high signal frequencies.[11]

3. Dynamic Conductivity Tensor

The ac in-plane components of the s-CNTs contribution to the conductivity tensor, $\langle \sigma_\omega^s \rangle_{xx}$ and $\langle \sigma_\omega^s \rangle_{yy}$, of the CNT network averaged over the CNT angles $\langle \sigma_\omega \rangle$ can be presented as

$$\langle \sigma_\omega^s \rangle_{xx} = \frac{2eN_s}{\pi \hbar} \int_{-\pi/2}^{\pi/2} d\alpha \, \cos(\alpha) \, \Theta(\alpha) \int_{-\infty}^{\infty} \frac{dpp}{m_s} \delta f^s(p,\alpha), \tag{2}$$

$$\langle \sigma_\omega^s \rangle_{yy} = \frac{2eN_s}{\pi \hbar} \int_{-\pi/2}^{\pi/2} d\alpha \, \sin(\alpha) \, \Theta(\alpha) \int_{-\infty}^{\infty} \frac{dpp}{m_s} \delta f^s(p,\alpha). \tag{3}$$

Here $\hbar$ is the Planck constant and $\Theta(\alpha)$ is the distribution function of the CNTs over their direction angle α. Using Eq. (1), from Eqs. (2) and (3) we obtain

$$\langle \sigma_\omega^s \rangle_{xx} = \theta_{xx} \frac{2ie^2 N_s}{\pi \hbar \, m_s} \int_{-\infty}^{\infty} \frac{dpp}{(\omega + i\nu_s)} \left[-\frac{df_0^s(p)}{dp} \right], \tag{4}$$

$$\langle \sigma_\omega^s \rangle_{yy} = \theta_{yy} \frac{2ie^2 N_s}{\pi \hbar \, m_s} \int_{-\infty}^{\infty} \frac{dpp}{(\omega + i\nu_s)} \left[-\frac{df_0^s(p)}{dp} \right], \tag{5}$$

where $\theta_{xx} = \int_{-\pi/2}^{\pi/2} d\alpha \, \cos^2(\alpha) \, \Theta(\alpha)$ and $\theta_{yy} = \int_{-\pi/2}^{\pi/2} d\alpha \, \sin^2(\alpha) \, \Theta(\alpha)$. In particular, the angle distribution parameters $\theta_{xx} = 1$ and $\theta_{yy} = 0$ correspond to the case when all the CNTs are aligned perpendicular to the electrodes. If the CNTs are uniformly distributed over the angles, $\theta_{xx} = \theta_{yy} = 1/2$.

When the 1DHSs in each s-CNT and m-CNT is degenerate, i.e., the hole Fermi energy in the s-CNT $\mu_s = \mu - \Delta_g/2$ markedly exceeds the temperature T, where the Fermi energy μ is counted from the Dirac point, from Eqs. (4) an (5) we obtain

$$\langle \sigma_\omega^s \rangle_{xx} \simeq \frac{i4e^2 N_s}{\pi \hbar} \sqrt{\frac{2\mu_s}{m_s}} \frac{\theta_{xx}}{(\omega + i\nu_s)}, \tag{6}$$

$$\langle \sigma_\omega^s \rangle_{yy} \simeq \frac{i4e^2 N_s}{\pi \hbar} \sqrt{\frac{2\mu_s}{m_s}} \frac{\theta_{yy}}{(\omega + i\nu_s)}. \tag{7}$$

Considering that the dc surface hole density in the s-CNTs Σ_0^s is given by the following formula:

$$\Sigma_0^s = \frac{4N_s}{2\pi \hbar} \int_{-\infty}^{\infty} dp f_0^s(p) \simeq \frac{4N_s \sqrt{2m_s \mu_s}}{\pi \hbar}. \tag{8}$$

Equations (6) and (7) can be presented as

$$\langle \sigma_\omega^s \rangle_{xx} \simeq \frac{ie^2 \Sigma_0^s}{m_s(\omega + i\nu_s)} \theta_{xx}, \qquad \langle \sigma_\omega^s \rangle_{yy} \simeq \frac{ie^2 \Sigma_0^s}{m_s(\omega + i\nu_s)} \theta_{yy}. \tag{9}$$

The ac conductivity associated with the m-CNTs of the density N_m can be calculated using the following expressions:

$$\langle \sigma_\omega^m \rangle_{xx} = \frac{4eN_m v_W}{\pi \hbar} \int_{-\pi/2}^{\pi/2} d\alpha \, \cos(\alpha) \, \Theta(\alpha) \int_0^\infty dp \, \delta f^m(p, \alpha), \tag{10}$$

$$\langle \sigma_\omega^m \rangle_{yy} = \frac{4eN_m v_W}{\pi \hbar} \int_{-\pi/2}^{\pi/2} d\alpha \, \sin(\alpha) \, \Theta(\alpha) \int_0^\infty dp \, \delta f^m(p, \alpha). \tag{11}$$

Accounting for Eq. (1), Eqs. (10) and (11) yield

$$\langle \sigma_\omega^m \rangle_{xx} = \frac{i4e^2 N_m}{\pi \hbar} \frac{v_W}{(\omega + i\nu)} \frac{\theta_{xx}}{[1 + \exp(-\mu_m/T)]} \simeq \frac{i4e^2 N_m}{\pi \hbar} \frac{v_W}{(\omega + i\nu_m)} \theta_{xx}, \tag{12}$$

$$\langle \sigma_\omega^m \rangle_{yy} = \frac{i4e^2 N_m}{\pi \hbar} \frac{v_W}{(\omega + i\nu)} \frac{\theta_{yy}}{[1 + \exp(-\mu_m/T)]} \simeq \frac{i4e^2 N_m}{\pi \hbar} \frac{v_W}{(\omega + i\nu_m)} \theta_{yy}. \tag{13}$$

Here $\mu_m = \mu \gg T$. Simultaneously, using the relation between the hole density in the m-CNTs Σ_0^m and μ_m one can obtain [compare with Eqs. (9)]:

$$\langle \sigma_\omega^m \rangle_{xx} \simeq \frac{ie^2 \Sigma_0^m}{m_m(\omega + i\nu_m)} \theta_{xx}, \qquad \langle \sigma_\omega^m \rangle_{yy} \simeq \frac{ie^2 \Sigma_0^m}{m_m(\omega + i\nu_m)} \theta_{yy}. \tag{14}$$

The quantity $m_m = \mu_m / v_W^2 = \mu / v_W^2$ in degenerate electron systems with a linear dispersion is usually called as the fictitious mass.

Comparing Eqs. (6) and (7) with Eqs. (10) and (11), we find

$$r = \frac{\langle \sigma_\omega^s \rangle_{xx}}{\langle \sigma_\omega^m \rangle_{xx}} = \frac{\langle \sigma_\omega^s \rangle_{yy}}{\langle \sigma_\omega^m \rangle_{yy}} = \frac{\sqrt{2\mu_s/m_s}}{v_W} \left(\frac{\omega + i\nu_m}{\omega + i\nu_s} \right) \frac{N_s}{N_m}. \tag{15}$$

This equation generalizes that obtained previously.[18] In the case of low $\omega \ll \nu_s, \nu_m$ and high signal frequencies $\omega \gg \nu_s, \nu_m$, assuming that ν_s and ν_m are proportional to the density of states at Fermi level in the pertinent CNTs, from Eq. (15) one can find, respectively

$$r_0 \simeq \frac{2\mu_s}{m_s v_W^2} \frac{N_s}{N_m}, \qquad r_\infty \simeq \sqrt{\frac{2\mu_s}{m_s v_W^2} \frac{N_s}{N_m}}. \tag{16}$$

For $N_s/N_m = 2$, $m_s = 6 \times 10^{-29}$ g, and $\mu_s = 30 - 60$ meV, we obtain $r_0 \simeq 0.375 - 0.650$ and and $r_\infty \simeq 0.80 - 1.13$.

4. 2D Plasmons in the CNT Networks

When the spacing between the CNT network plane and the gate W_g is small in comparison with the distance between electrodes L and the plasma wave length λ, the self-consistent ac potential at the CNT network plane $\delta\varphi_\omega = \delta\varphi_\omega(x, y, 0)$ can be calculated from the Poisson equation in the gradual channel approximation:[20]

$$\frac{\delta\varphi_\omega}{W_g} = \frac{4\pi e}{\ae}\delta\Sigma_\omega. \tag{17}$$

Here $\delta\Sigma_\omega = \delta\Sigma_\omega^s + \delta\Sigma_\omega^m$ is the ac component of the net hole density in the CNT network, $\ae$ is the dielectric constant of the gate layer material. The axis z is perpendicular to the network plane.

For simplicity, in Eq. (11) we omitted the factor associated with the quantum capacitance,[21] because for the practical values of the gate layer thickness it is close to unity. This implies that we assume the geometrical capacitance per unit area $C_g = \ae/4\pi W_g$ to be much smaller than the quantum capacitance.

The continuity equations of the hole components in the s- and m-CNTs are as follows:

$$-i\omega e\delta\Sigma_\omega^s + \langle\sigma_\omega^s\rangle_{xx}\frac{d^2\delta\varphi_\omega}{dx^2} + \langle\sigma_\omega^s\rangle_{yy}\frac{d^2\delta\varphi_\omega}{dy^2} = 0, \tag{18}$$

$$-i\omega e\delta\Sigma_\omega^m + \langle\sigma_\omega^m\rangle_{xx}\frac{d^2\delta\varphi_\omega}{dx^2} + \langle\sigma_\omega^m\rangle_{yy}\frac{d^2\delta\varphi_\omega}{dy^2} = 0. \tag{19}$$

For the ac component of the self-consistent potential, which is searched in the form of the wave propagating along the axis y, i.e., in the form $\delta\varphi_\omega = \delta\psi(x)\exp[i(ky - \omega t)]$, Eqs. (17)–(19) lead to the following equation:

$$\theta_{xx}\frac{d^2\delta\varphi_\omega}{dx^2} + \left[\frac{\pi^2\omega(\omega + i\nu)}{\omega_p^2} - \theta_{yy}k^2L^2\right]\frac{\delta\varphi_\omega}{L^2} = 0. \tag{20}$$

If $\nu_s = \nu_m = \nu$, we obtain the following formulas for the characteristic plasmon frequency ω_p and velocity s:

$$\omega_p = \frac{\pi s}{L}, \qquad s = \sqrt{\frac{4\pi e^2 W_g}{\ae}\left(\frac{\Sigma_0^s}{m_s} + \frac{\Sigma_0^m}{m_m}\right)}. \tag{21}$$

At $W_g = 10^{-5}$ cm, $\ae = 4$, $m_s = 6 \times 10^{-29}$ g, $\Sigma_0^s + \Sigma_0^m(m_s/m_n) = 5 \times 10^{11}$ cm^{-1}, one obtains $s \simeq 2.5 \times 10^8$ cm/s. If $L = (0.5 - 1.5)$ μm, the characteristic plasmon frequency falls into the THz range: $\omega_p/2\pi \simeq (0.83 - 2.5)$ THz.

Equations (18)–(21) differ from the related equations obtained previously[18] by describing the plasmons with finite wave number k characterizing their propagation along the strip electrodes.

For the case of connected ohmic electrodes, the boundary conditions for Eq. (20) have the following form:

$$\delta\varphi_\omega|_{x=0} = 0, \qquad \delta\varphi_\omega|_{x=L} = 0. \tag{22}$$

Solving Eq. (20) with boundary conditions (22), we arrive at the following dispersion relation for the plasmons propagating along the slot line with the CNT network:

$$\omega(\omega + i\nu) = s^2 \left(\frac{\pi^2 n^2}{L^2} \theta_{xx} + k^2 \theta_{yy} \right), \tag{23}$$

where $n = 1, 2, 3, \ldots$ is the index of the plasmonic mode. Equation (23) yields

$$\text{Re } \omega = \sqrt{s^2 \left(\frac{\pi^2 n^2}{L^2} \theta_{xx} + k^2 \theta_{yy} \right) - \frac{\nu^2}{4}}. \tag{24}$$

Due to the dependence of the plasmon velocity s on the hole densities in the s- and m-CNTs on the gate voltage V_g, the spectrum of the plasmons can be controlled by this voltage. It is instructive that the s versus V_g relation differs from those in the gated structures with the parabolic energy spectrum in the gated graphene structures. If the hole density is determined solely by the holes induced by the gate voltage V_g, one obtaines $s \propto V_g^{1/2}$ in the former structures[2] and $s \propto V_g^{1/4}$ in the gated graphene structures.[22] This is because the CNT network under consideration consists of a mixture of different CNTs. Naturally, the variation of N_s/N_m affects the s versus V_g relation.

As seen from Eq. (24), the plasmon spectrum is sensitive to the angular distributions of the CNTs. Figures 2 and 3 show the dispersion relations for the plasmon modes in the CNT network structures with different spacing between the electrodes L, different parameters θ_{xx} and θ_{yy}, characterizing the angular distributions of the CNTs, and different collision frequency ν.

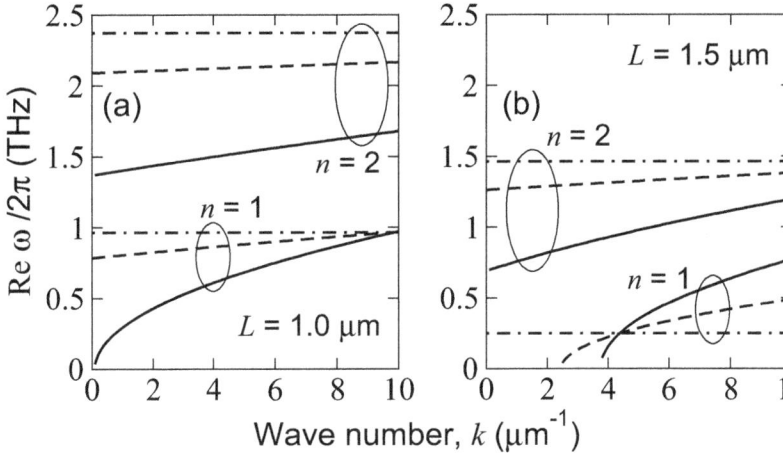

Fig. 2. Real parts of the plasmon frequency Re ω versus wave number k for different modes in the CNT networks with (a) $L = 1.0 \times 10^{-4}$ cm and (b) $L = 1.5 \times 10^{-4}$ cm and different θ_{xx} and θ_{yy}: dash-dotted lines - $\theta_{xx} = 1$, $\theta_{yy} = 0$, dashed lines - $\theta_{xx} = 0.8$, $\theta_{yy} = 0.2$, solid lines - $\theta_{xx} = 0.4$, $\theta_{yy} = 0.6$ ($s = 2.5 \times 10^8$ cm/s and $\nu = 10 \times 10^{12}$ s^{-1}).

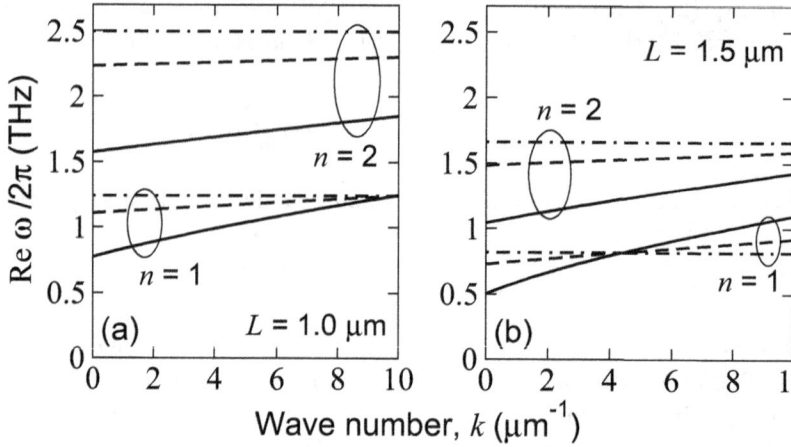

Fig. 3. The same as in Fig. 2 but for $\nu = 2 \times 10^{12}$ s^{-1}.

As follows from Figs. 2 and 3, the spectrum of the plasmons, in particular, their limiting frequencies Re $\omega/2\pi|_{k=0}$ and the Re ω versus k relations substantially vary with varying angular distribution of the CNTs. The limiting plasmon frequency Re $\omega/2\pi|_{k=0}$ markedly drops when θ_{xx} decreases. Moreover, at relatively small θ_{xx}, the plasmonic modes can exist only when k is sufficiently large.

5. Conclusions

We calculated the dynamic conductivity of the gated lateral CNT networks (CNT felt) as functions of the densities of carriers (holes) in the s- and m-CNTs and the CNT angular distribution. We also obtained the dispersion relations for the plasmons propagating in the lateral CNT networks along the highly conducting strips forming a slot line. As shown, the plasmon spectrum substantially depends not only on the carrier density and the gate voltage but also on the CNT angular distribution. For the realistic parameters, the plasmon frequency in the lateral CNT networks can be in the THz range. Such CNT network structures can be used as passive and active elements in a variety of THz devices.

Acknowledgments

The work was supported by the Japan Society for Promotion of Science (Grant-in-Aid for Specially Promoted Research No. 23000008 and Grant No. 16H06361), the Russian Scientific Foundation (Grants No. 14-29-00277 and No. 1619-10557), the Russian Foundation for Basic Research (Grant No. 16-29-03033), and the Ministry of Education and Science of the Russian Federation (Grant No. 16-19-2014/K and Contract No. 14.B25.31.0007). The works at RPI was supported by the US Army Research Laboratory Cooperative Research Agreement.

References

1. M. Dyakonov and M. Shur, "Shallow water analogy for a ballistic field effect transistor: New mechanism of plasma wave generation by dc current", *Phys. Rev. Lett.* **71**(15) (1993) 2465–2468.

2. M. I. Dyakonov and M. S. Shur, "Plasma wave electronics: novel terahertz devices using two-dimensional electron fluid", *IEEE Trans. Electron Devices* **43** (1996) 1640–1645.

3. H. M. Manohara, E. W. Wong, E. Schlecht, B. D. Hunt, and P. H. Siegel, "Carbon nanotube Schottky diodes using Ti-Schottky and Pt-Ohmic contacts for high frequency applications", *Nano Lett.* **5**(7) (2005) 1469–1474.

4. J. D. Chudow, D. F. Santavicca, C. B. McKitterick, D. E. Prober, and P. Kim, "Terahertz detection mechanism and contact capacitance of individual metallic single-walled carbon nanotubes", *Appl. Phys. Lett.* **100**(16) (2012) 163503.

5. G. Fedorov, A. Kardakova, I. Gayduchenko, I. Charayev, B. M. Voronov, M. Finkel, T. M. Klapwijk, S. Morozov, M. Presniakov, I. Bobrinetskiy, R. Ibragimov, and G. Goltsman, "Photothermoelectric response in asymmetric carbon nanotube devices exposed to sub-terahertz radiation", *Appl. Phys. Lett.* **103**(18) (2013) 181121.

6. X. He, N. Fujimura, J. M. Lloyd, K. J. Erickson, A. A. Talin, Q. Zhang, W. Gao, Q. Jiang, Y. Kawano, R. H. Hauge, F. Léonard, and J. Kono, "Carbon nanotube terahertz detector", *Nano Lett.* **14**(7) (2014) 3953–3958.

7. I. Gayduchenko, A. Kardakova, G. Fedorov, B. Voronov, M. Finkel, D. Jimenez, S. Morozov, M. Presniakov, and G. Goltzman, "Response of asymmetric carbon nanotube network devices to sub-terahertz and terahertz radiation", *J. Appl. Phys.* **118** (2015) 194303.

8. X. He, F. Léonard, and J. Kono, "Uncooled Carbon nanotube photodetectors", *Advanced Optical Materials* **3** (2015) 989–1011.

9. H. E. Ruda and A. Shik, "Polarization-sensitive optical phenomena in semiconducting and metallic nanowires", *Phys. Rev. B* **72**(11) (2005) 115308.

10. G. Ya. Slepyan, M. V. Shuba, and S. A. Maksimenko, "Theory of optical scattering by achiral carbon nanotubes and their potential as optical nanoantennas", *Phys. Rev. B* **73**(19) (2006) 195416.

11. A. P. Dmitriev and M. S. Shur, "One dimensional plasmons in pyroelectric-semiconductor composites", *J. Appl. Phys.* **103**(8) (2008) 084511.

12. T. Nakanishi and T. Ando, "Optical response of finite-length carbon nanotubes", *J. Phys. Soc. Japan* **78** (2009) 114708.

13. G. Ya. Slepyan, M. V. Shuba, and S. A. Maksimenko, C. Thomsen, and A. Lakhtakia, "Terahertz conductivity peak in composite materials containing carbon nanotubes: Theory and interpretation of experiment", *Phys. Rev. B* **81**(20) (2010) 205423.

14. Qi Zhang, E. H. Haroz, Z. Jin, L. Ren, X. Wang, R. S. Arvidson, A. Luttge, and J. Kono, "Plasmonic nature of the terahertz conductivity peak in single-wall carbon nanotubes", *Nano Lett.* **13**(12) (2013) 5991–5996.

15. X. He, W. Gao, Qi Zhang, L. Ren, and J. Kono, "Carbon-based THz devices", *Proc. SPIE* **9476** (2015) 947612.

16. T. Morimoto, S.-K. Joung, T. Saito, D. N. Futaba, K. Hata, and T. Okazaki, "Length-dependent plasmon resonance in single-walled carbon nanotubes", *ACS Nano* **8**(10) (2014) 9897–9904.

17. Zh. Shi, X. Hong, H. A. Bechtel, Bo Zeng, M. C. Martin, K. Watanabe, T. Taniguchi, Y.-R. Shen, and F. Wang, "Observation of a Luttinger-liquid plasmon in metallic single-walled carbon nanotubes", *Nat. Photonics* **9** (2015) 515–519.

18. V. V. Ryzhii, T. Otsuji, M. Ryzhii, V. G. Leiman, G. Fedorov, G. N. Goltzman, I. A. Gayduchenko, N. Titova, D. Coquillat, D. But, W. Knap, V. Mitin, and M. S. Shur, "Two-dimensional plasmons in lateral carbon nanotube network structures and their effect on the terahertz radiation detection", *J. Appl. Phys.* **120**(4) (2016) 044501.

19. R. Saito, G. Dresselhaus, and M. S. Dresselhaus, *Physical Properties of Carbon Nanotubes*, Imperial College Press, London, 1998.

20. M. S. Shur, *Physics of Semiconductor Devices*, Prentice Hall, New Jersey, 1990.

21. S. Luryi, "Quantum capacitance devices", *Appl. Phys. Lett.* **52**(6) (1988) 501–503.

22. V. Ryzhii, A. Satou, and T. Otsuji, "Plasma waves in two-dimensional electron-hole system in gated graphene heterostructures", *J. Appl. Phys.* **101** (2007) 024509.

Integrated On-Chip Nano-Optomechanical Systems

Zhu Diao*

*Department of Mathematics, Physics and Electrical Engineering,
Halmstad University, Box 823, SE-301 18 Halmstad, Sweden
diaoz@tcd.ie*

Vincent T. K. Sauer

*National Institute for Nanotechnology, 11421 Saskatchewan Drive,
Edmonton T6G 2M9, Alberta, Canada
Department of Electrical and Computer Engineering, University of Alberta,
9211-116 Street NW, Edmonton T6G 1H9, Alberta, Canada
and
Department of Biological Sciences, CW 405, Biological Sciences Bldg.,
University of Alberta, Edmonton T6G 2E9, Alberta, Canada
vsauer@ualberta.ca*

Wayne K. Hiebert

*National Institute for Nanotechnology, 11421 Saskatchewan Drive,
Edmonton T6G 2M9, Alberta, Canada
and
Department of Physics, University of Alberta, 4-181 CCIS,
Edmonton T6G 2E1, Alberta, Canada
wayne.hiebert@nrc-cnrc.gc.ca*

Recent developments in integrated on-chip nano-optomechanical systems are reviewed. Silicon-based nano-optomechanical devices are fabricated by a two-step process, where the first step is a foundry-enabled photonic circuits patterning and the second step involves in-house mechanical device release. We show theoretically that the enhanced responsivity of near-field optical transduction of mechanical displacement in on-chip nano-optomechanical systems originates from the finesse of the optical cavity to which the mechanical device couples. An enhancement in responsivity of more than two orders of magnitude has been observed when compared side-by-side with free-space interferometry readout. We further demonstrate two approaches to facilitate large-scale device integration, namely, wavelength-division multiplexing and frequency-division

*Previous address: National Institute for Nanotechnology, 11421 Saskatchewan Drive, Edmonton T6G 2M9, Alberta, Canada and Department of Physics, University of Alberta, 4-181 CCIS, Edmonton T6G 2E1, Alberta, Canada.

multiplexing. They are capable of significantly simplifying the design complexity for addressing individual nano-optomechanical devices embedded in a large array.

Keywords: Nanomechanical resonators; Nano-optomechanical systems; Nanophotonics.

1. Introduction

Rapid developments in nanotechnology and nanoelectronics have made it possible to routinely design and fabricate devices and systems with functional blocks in the sub-micrometer scale. The current wave of "More than Moore" technology, which focuses on device functionalities rather than merely packing up more and more transistors into a smaller and smaller chip area, has led to the emergence of a wide range of innovative devices with their potential applications in remote sensing, point-of-care testing, and high-bandwidth signal processing.[1] Among them, microfabricated high-frequency nanomechanical resonators operated in the MHz - GHz frequency range have attracted a growing amount of attention.[2,3] These devices take full advantage of the flexibility and robustness offered by precisely-tuned nanometer-scale mechanical structures, and have been utilized as ultra-sensitive surface force gauges,[4,5] inertial mass sensors,[6,7,8,9] magnetometers,[10,11,12,13,14] thermometers,[15,16] memory cells,[17] accelerometers,[18] and clocking references.[19]

One of the factors which so far have limited further developments and wide adoption of nanomechanical resonators is how the diminutive displacement involved while operating these resonators (typically in the sub-nanometer scale) may be effectively transduced with a GHz-plus bandwidth. In nanoelectromechanical systems (NEMS), the mechanical displacement is actuated as well as sensed electrically, *i.e.* through a selection of capacitive,[20,21,22] piezoelectric,[23,24] piezoresistive,[25,26,27] electron tunnelling,[28,29] and electrothermal[30] methods or a combination of several of them together. These tranduction schemes often rely on on-chip integrated signal amplifiers, *e.g.* single-electron transistors (SET)[31,32] or high-electron-mobility transistors (HEMT)[33,34] to enhance the relatively small electrical signal induced by device displacement, and complicated down-mixing circuits to bypass the bandwidth limitation.[22,26,27,29,30] Such approaches drastically increase the design complexity and very often limit the operating temperatures and applicable areas of NEMS.

Free-space optical methods have long been utilized to transduce the motion of nanomechanical resonators.[35,36,37,38,39] Compared with electrical transduction, high-bandwidth operation is easily achieved without signal down-mixing. However, due to natural constrains of the optical wavelength, upon scaling down, physical dimensions of nanomechanical resonators quickly reach the diffraction limit of the laser wavelength in use.[40] Most of the input laser beam then undergoes a diffraction process rather than reflection, strongly reducing the intensity of light that can be captured on the photodetector. Hence, the signal-to-noise ratio of the readout quickly diminishes. Furthermore, stringent requirements on the relative alignment of the laser spot and the nanomechanical device hinders large-scale device integration. Nano-optomechanical systems (NOMS), where on-chip integrated optical circuits are utilized to transduce the motion of nanomechanical resonators in the optical near-field, form a significant development in the last decade.[41,42,43,44,45] It overcomes the diffraction limit by moving the interaction from far-field to near-field, while maintaining the high bandwidth offered by an optical transduction scheme. Taking advantage of high-finesse on-chip optical cavities, where individual photons can interact multiple times with a nanomechanical device before they are dissipated, the responsivity of NOMS displacement transduction has reached an unprecedented level in recent years.[46,47,48,49] It was further demonstrated that the optical gradient force existing in the near-field of a guided light wave can be harnessed to excite the nanomechanical device, making possible full-optical pump-probe operation.[50,51,52,53] Moreover, developments in NOMS coincide with the rapid advancement of silicon-based nanophotonics technology,[54] providing a clear pathway for chip-level device integration.

Driving forces behind the rapid expansion of NOMS research in the last decade have been and are still of two folds. One is the pursuing of a deep understanding of the quantum nature of our universe, in which researchers are trying to create macroscopic quantum mechanical systems by realizing strong coupling between mechanical resonators and cavity fields (the so-called cavity optomechanics).[55,56,57,58,59,60] The other is the demand of the microelectronics and nanophotonics industry to develop mechanically-active functional devices in the nanometer scale and establish highly-sensitive, reliable, low-cost, highly-integrable transduction schemes to monitor their displacement.[41,43,45,51,52,61] Hence, NOMS is a perfect modern-day example where cutting-edge fundamental science meets innovative industry development. While we emphasize that the two aspects are both of paramount importance in the advancement of NOMS technology, in our current report, we will limit ourselves mostly to the later aspect.

Here we review recent developments in high-frequency integrated NOMS realized on the silicon-on-insulator platform. These devices are compatible with state-of-the-art silicon photonics technology, thus, hold the promise to be implemented imminently in a large variety of applications as sensors, actuators, and frequency references. In Section 2, we outline the theoretical background of NOMS transduction in the linear regime and in Section 3, the device fabrication process is explained. Our discussion here pays particular attention to the fact that integrated NOMS can be mass manufactured in a similar process as that used to fabricate silicon photonic circuits. Hence, it is especially suitable for large-scale device integration. Measurement results on a selective subset of our NOMS devices are presented in Section 4, with a focus on how NOMS technology may move from its current state of addressing individual devices to chip-level multiplexing and multi-scale integration, before conclusions are made in Section 5.

2. Theory of Device Displacement Transduction

A comprehensive theoretical description of the problem in which a nanome-chanical resonator couples to an optical cavity requires the language of cavity optomechanics.[57,59,60] It not only takes account of the effective refractive index change of the cavity as a result of the displacement of the mechanical object, but also a 'back-action' exerted by the cavity to it. However, when the finesse of the optical cavity is low, the opto-mechanical coupling is relatively weak, and the me-chanical displacement in concern is small, the discussion may be simplified to the linear regime, where the effect of the 'back-action' can be neglected. This is the approach we adopt. We will show later in this section that such assumptions hold in our NOMS devices and this simplified theoretical treatment is supported by our experimental data.

From now on, we will focus our discussion on one of the simplest cases in which an optical ring or race-track cavity is coupled to a bus waveguide in the all-pass filter configuration on one side and a released cantilever or a doubly-clamped mechanical beam is coupled to the same cavity on the opposite side. A sketch of such a device is shown in Fig. 1(a). Light is coupled into the photonic circuit through grating couplers.[62] The bus waveguide carries light to the race-track optical cavity and also carries light away from it to the output grating coupler. The transmission coefficient of such a device goes through a series of dips when the wavelength of the incoming light satisfies the cavity resonance condition as shown in Fig. 1(b). More specifically, the transmission coefficient can be expressed as[63,64,65]

$$T = \frac{T_0 + [(2/\pi)\mathcal{F} \cdot \sin(\phi/2)]^2}{1 + [(2/\pi)\mathcal{F} \cdot \sin(\phi/2)]^2}. \tag{1}$$

where T_0 is the on-resonance residual transmission determined by the coupling between the cavity and the waveguide, $\phi = (\frac{2\pi}{\lambda})n_{\text{eff}}L$ is the round-trip phase accu-mulation, in which λ denotes the vacuum wavelength, n_{eff} is the effective refractive index and L is the round-trip distance, and $\mathcal{F} = \text{FSR}/(2\delta\lambda)$ is the finesse of the optical cavity. The (wavelength) free-spectral range, FSR, is the separation in wave-length of adjacent minima in the transmission coefficient and $2\delta\lambda$ is the full-width-at-half-maximum of a specific resonance mode.[66] A parameter closely-related to the FSR is the optical cavity quality factor, $Q_{\text{o}} = (n_{\text{eff}}L/\lambda)\mathcal{F}$. Finesse is governed by the round-trip cavity loss and does not directly depend on the length of the cavity, whereas Q_{o} is a function of both parameters. We will see later that the efficiency of displacement transduction in NOMS depends solely on the finesse of the cavity, hence, the transduction responsivity cannot be enhanced by merely increasing the length of the cavity to obtain apparently 'sharp' cavity resonances while plotted as a function of wavelength.

Displacement of the mechanical resonator (denoted as x) tunes the effective re-fractive index of the section of the optical cavity coupled to it, as shown in Fig. 2. When the mechanical resonator moves closer to the race-track optical cavity, the ef-fective refractive index increases. The slope expressed as $\partial n_{\text{eff}}/\partial x$ can be extremely large as seen in Fig. 2(b) when the gap between the mechanical resonator and the cavity is small. The refractive index change in turn modifies the round-trip phase accumulation, the position of the cavity mode and the transmission coefficient of the system. One can write[64,65]

$$\frac{\partial T}{\partial x} = \frac{\partial T}{\partial \phi} \frac{\partial \phi}{\partial x}.$$
(2)

It is important to note that the first term on the right hand side of Eq. (2) is an intrinsic property of the all-pass optical filter consisted of the optical cavity and the input bus waveguide while the second term describes the strength of coupling between the mechanical resonator and the cavity. To achieve high transduction responsivity, one has to optimize $\partial T/\partial x$.

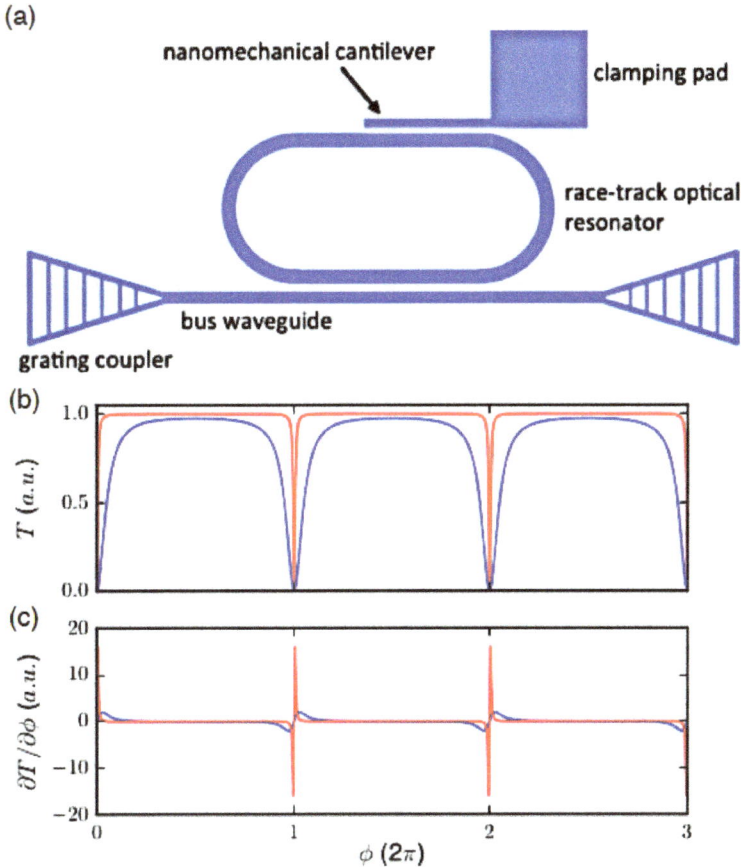

Fig. 1. (a) A sketch showing a nanomechanical cantilever coupling to an on-chip all-pass filter. The components are not drawn to scale. (b) The transmission coefficient of an all-pass filter with $\mathcal{F} = 10$ (blue) and $\mathcal{F} = 100$ (red). (c) The derivative of the transmission coefficient, $\partial T/\partial \phi$, for $\mathcal{F} = 10$ (blue) and $\mathcal{F} = 100$. T_0 in Eq.(1) is set to zero in (b) and (c).

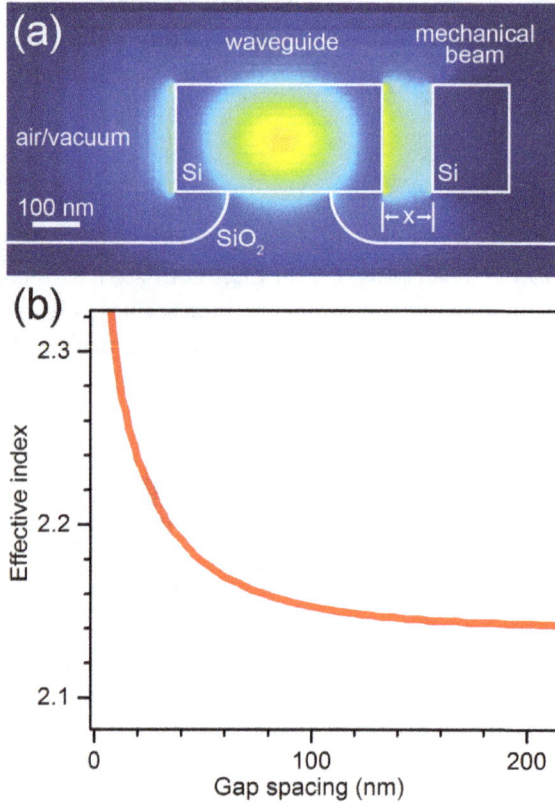

Fig. 2. (a) The simulated optical mode in a partially underetched waveguide adjacent to a released mechanical beam. The waveguide is 430 nm wide and the width of the mechanical beam is 160 nm. The distance between them is 100 nm. The waveguide forms part of a race-track optical cavity. (b) The waveguide effective refractive index as a function of its distance to the mechanical resonator.

Maxima in $|\partial T/\partial \phi|$ occur when $\partial^2 T/\partial \phi^2$ goes to zero.[64,65]

$$\left|\frac{\partial T}{\partial \phi}\right|_{\max} \approx \frac{3\sqrt{3}}{8\pi}(1-T_0)\mathcal{F}. \tag{3}$$

and

$$\sin\left(\frac{\phi}{2}\right)\Bigg|_{|\frac{\partial T}{\partial \phi}| \text{ at max}} \approx \pm\frac{\pi}{2\sqrt{3}\mathcal{F}}. \tag{4}$$

Experimentally, using a tunable diode laser, one often sets the probe laser wavelength to where the transmission coefficient as a function of laser wavelength has the maximum slope, $|\partial T/\partial \lambda|_{\max}$. Strictly speaking, this is not the same condition as defined in Eqs. (3) and (4), since

$$\frac{\partial \phi}{\partial \lambda} = -\frac{2\pi n_{\text{eff}} L}{\lambda^2} + \left(\frac{2\pi L}{\lambda}\right)\left(\frac{\partial n_{\text{eff}}}{\partial \lambda}\right). \tag{5}$$

Only when the chromatic dispersion of the waveguide, $\partial n_{\text{eff}}/\partial\lambda$, can be ignored, may the two conditions converge. The second term on the right hand side in Eq. (2) can be further expanded as[65]

$$\frac{\partial\phi}{\partial x} = (\frac{2\pi}{\lambda}) \cdot (\frac{\partial n_{\text{eff}}}{\partial x}) \cdot \beta \cdot l. \tag{6}$$

in which l is the length of the nanomechanical device and β takes account into the mode shape of the mechanical resonance.[67]

Inserting Eqs. (3) and (6) into Eq. (2), one obtains the maximum transduction responsivity of a NOMS device

$$\left|\frac{\partial T}{\partial x}\right|_{\text{max}} = \frac{3\sqrt{3}}{8\pi} \cdot (1 - T_0) \cdot \mathcal{F} \cdot (\frac{2\pi}{\lambda}) \cdot (\frac{\partial n_{\text{eff}}}{\partial x}) \cdot \beta \cdot l. \tag{7}$$

We approximate the frequency response of the mechanical resonator as a high-Q harmonic oscillator, which can be described by a Lorentzian function.

$$S_{\text{x}}(f) = \frac{[F_0(f)/4\pi^2 m_{\text{eff}}]^2}{(f^2 - f_0^2)^2 + (\frac{ff_0}{Q_{\text{m}}})^2}. \tag{8}$$

in which $S_{\text{x}}(f)$ is the power spectral of the mechanical displacement, $F_0(f)$ is the magnitude of the frequency-dependent driving force, m_{eff} is the effective mass of the resonator, and f_0 is the mechanical resonance frequency. At a finite temperature with no apparent external driving force, the thermal bath provides a wide-band excitation to a mechanical device and its response is governed by Eq. (8), the so-called thermomechanical (TM) noise. Since the system has only a single degree of freedom, its TM noise can be calibrated utilizing the equipartition theorem

$$\frac{1}{2}k_{\text{B}}T = \frac{1}{2}k\langle x^2\rangle. \tag{9}$$

where k_{B} is the Boltzmann constant, and $k = (2\pi f_0)^2 m_{\text{eff}}$ is the effective spring constant of the mechanical resonator. This provides a unique way to confirm the validity of Eq. (7), since the TM displacement is known from Eq. (9), $\partial T/\partial\lambda$ and $\mathcal{F}$ can be obtained from the device DC transmission spectrum, β and l are both geometrical parameters, and $\partial n_{\text{eff}}/\partial x$ can be worked out through simulations. Such validity tests have been carried out with a series of nanomechanical cantilevers of different lengths and placed at different distances away from the race-track optical cavity in Ref. 65, and the results are shown in Fig. 3. For this series of devices, the finesse of the race-track optical resonators is between 50 and 70, whereas the nanomechanical cantilevers are placed 90 - 160 nm away from the optical cavities. The close correspondence between the theory and the experiment demonstrates that our theoretical framework is sufficient to describe the responsivity of NOMS devices in the low-finesse, low-optomechanical-coupling limit where the cavity 'back-action' is weak.

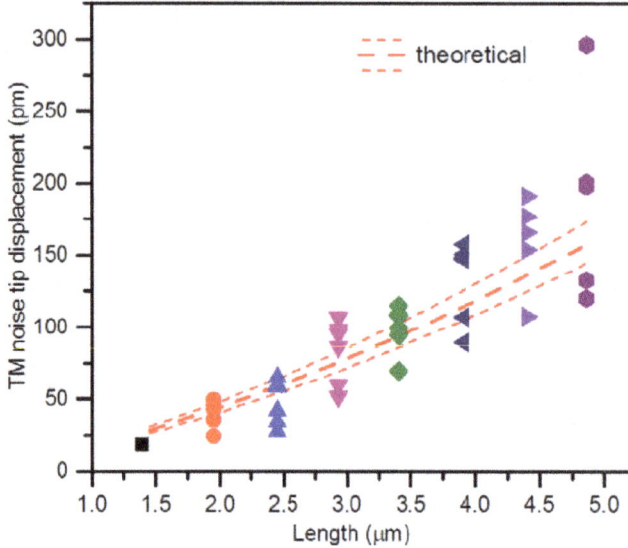

Fig. 3. The cantilever tip displacement caused by thermomechanical noise. The different data points for each length come from devices with varying measured gap spacing. The spread in the data for each length is due to the slight variation in resonator widths with different gap sizes due to proximity effects during fabrication. Smaller gaps pair with slightly thicker and stiffer beams. The red dashed lines are the calculated theoretical values for the average beam thickness of 160 ± 10 nm. Reprinted from V. T. K. Sauer, Z. Diao, M. R. Freeman, and W. K. Hiebert, "Optical racetrack resonator transduction of nanomechanical cantilevers", *Nanotechnology* **25** (2014) 055202 with permission from IOP publishing.

From Eq. (7), we see that the enhancement in transduction responsivity in NOMS readout, when compared with conventional free-space interferometry, originates largely from the finesse of the on-chip optical cavity. In our samples, the finesse of the optical cavity falls in the range of 20 - 150 while the Fabry-Pérot cavity formed by the top and bottom silicon layers as end 'mirrors' in SOI-based devices usually has a finesse in the order of 1. Thus, an improvement in the responsivity of the same order of magnitude may be expected. In Fig. 4, we show a side-by-side comparison of NOMS transduction and free-space interferometry transduction of the same nanomechanical doubly clamped beam, under the same excitation amplitude provided by a piezo disk.[68] The mechanical resonance mode is at $\sim$ 4.628 MHz and the quality factor $Q_m \sim 12,000$. NOMS transduction leads to a slightly lower resonance frequency which can be attributed to the optical spring effect.[69] The on-chip race-track cavity utilized to transduce the mechanical motion in the NOMS design has a finesse $\mathcal{F} \sim 134$ and the corresponding responsivity improvement is $\sim 150\times$ (the peak response is 0.8 mV for NOMS readout with a laser power of 47 μW and only 3 μV for free-space readout with a laser power of 28 μW), closely following the above discussion.

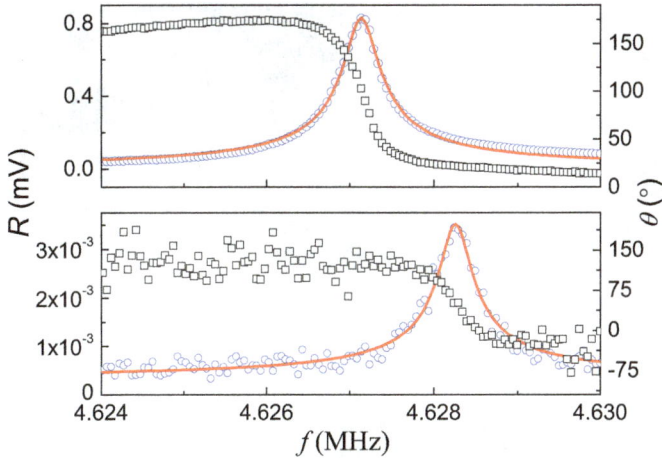

Fig. 4. Driven mechanical response of an $l = 15$ μm doubly clamped beam embedded in a race-track resonator measured by nanophotonic (top panel) and free-space interferometry readout (bottom panel). The optical power in/on the device is 47 μW for the NOMS readout whereas 28 μW for the free-space interferometry readout. Blue circles represent the amplitude signal while black squares represent the phase signal. Red lines are Lorentzian fits to the data. Adapted from Z. Diao, J. E. Losby, V. T. K. Sauer, J. N. Westwood, M. R. Freeman, and W. K. Hiebert, "Confocal scanner for highly sensitive photonic transduction of nanomechanical resonators", *Appl. Phys. Express* **6** (2013) 065202 with permission from the Japan Society of Applied Physics.

3. Device Fabrication

One of the major advantages of NOMS is that they can be fabricated in a similar process as that commonly used to produce state-of-the-art silicon-on-insulator (SOI) photonics, indicating a clear pathway for wafer-scale manufacturing. It also allows NOMS to leverage the vast readily available infrastructure in silicon technology and be fused seamlessly into conventional integrated circuit chips. We take full advantage of this fact by outsourcing the bulk of device fabrication work to silicon photonics foundries. More specifically, nanophotonic circuits presented in this work were fabricated by either LETI in Grenoble, France or IMEC in Leuven, Belgium through their multi-project wafer services. Their standard processes are both built upon 200 mm SOI wafers with 220 nm silicon device layers on top of 2 μm buried oxides (BOX). Wafers are exposed by either 248 nm or 193 nm deep UV lithography with a minimum achievable feature size of $\sim$ 120 nm. Centimeter-sized dies housing thousands of silicon photonic circuit designs are fabricated in a single run. Figure 5(a) is an optical image of a silicon photonic chip fabricated by IMEC while Fig. 5(b) – (d) show a number of zoom-in SEM images recorded upon various silicon nanophotonic components.

NOMS devices are then formed in a second in-house processing step during which predefined sections of the top silicon device layer are mobilized from the BOX using SiO_x wet etching, forming nanomechanical resonators. Resist layers patterned by

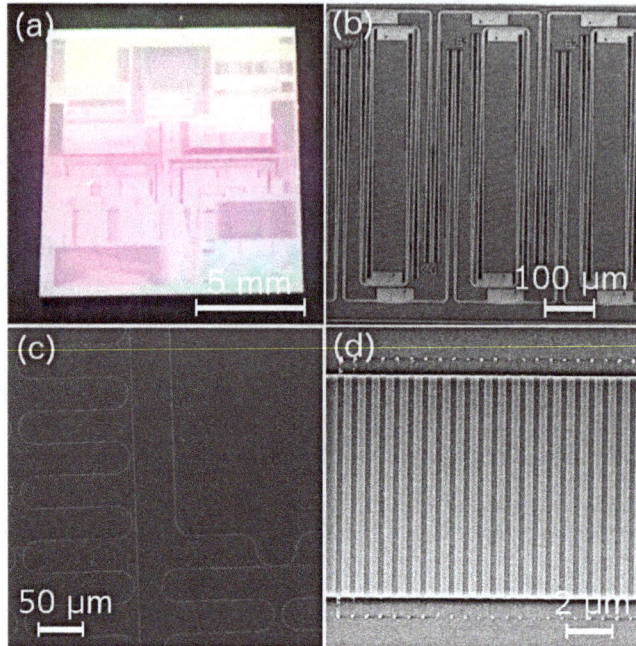

Fig. 5. (a) Full-view optical image of a silicon photonic chip containing a large number of NOMS devices fabricated by IMEC, Leuven, Belgium. (b) SEM image of a set of twelve adjacent photonic circuits on the same chip. (c) Zoomed-in view of photonic waveguides. (d) Zoomed-in view of a grating coupler used to couple light in and out of the photonic chip.

either UV lithography or electron beam lithography are used to protect the rest of the chip surface during the oxide etch. Alternatively, a timed wet etching process has also been developed to release nanomechanical devices with widths less than that of silicon photonic waveguides.[45] In Fig. 6, we present high-magnification SEM images of a number of completed NOMS devices. Resonance frequencies of our NOMS devices are designed to be in the range of several megahertz to several tens of megahertz, thus, in the high frequency (HF) band of the radio spectrum.

We note that our device fabrication process is of good reproducibility. Relatively little variation in device mechanical as well as optical properties can be identified in devices fabricated in the same processing run. In Fig. 7, we summarize the mechanical resonance frequency, f_0, and the mechanical quality factor, Q_m, of a number of NOMS doubly clamped beams with nominally the same physical dimensions. One can easily see that variations in both f_0 and Q_m between device and device are fairly small, with the standard deviation in f_0 being less than 2% of $\langle f_0 \rangle$. Although the standard deviation in Q_m is larger, it still only amounts to 18% of $\langle Q_m \rangle$. Further improvement in the conformity of the device mechanical quality factor may be gained through fine control of clamping losses in the device design.[70,71]

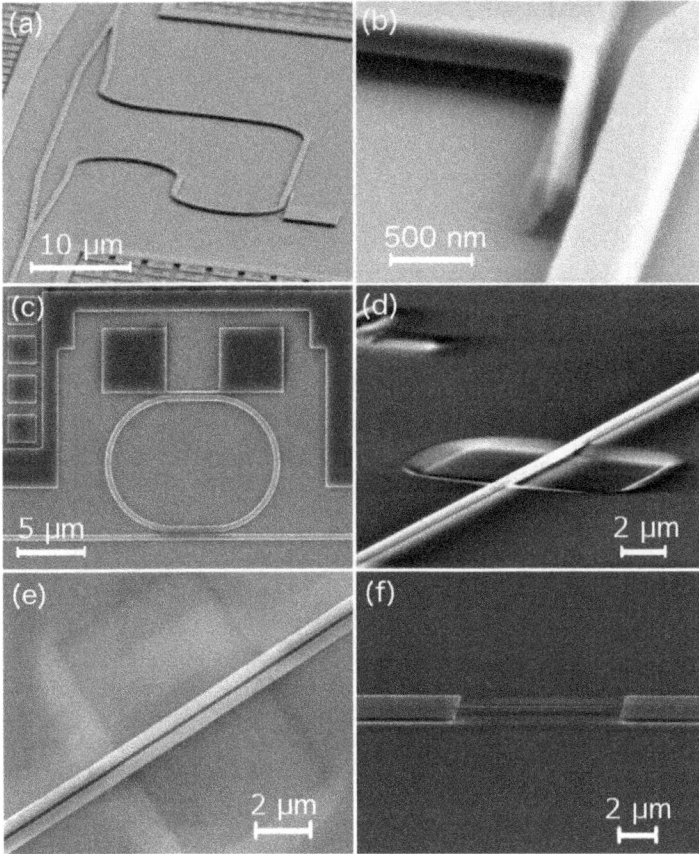

Fig. 6. SEM images of released NOMS devices. (a) A cantilever coupled to an on-chip Mach-Zehnder interferometer. (b) The same device as in (a) under high magnification. Electron beam charging causes the cantilever to vibrate during imaging. (c) A doubly clamped beam coupled to a race-track optical cavity. (d) A doubly clamped beam coupled to the partially etched BOX layer. (e) Double NOMS beams. (f) A doubly clamped beam with clamping pads.

4. Measurement Results

NOMS devices are housed in a vacuum chamber which can be pumped down to $< 1 \times 10^{-5}$ Torr to eliminate air damping. The custom designed measurement setup was previously described in Ref. 68, where one and the same microscope objective is utilized to both couple light into the photonic chip and retrieve light coming out of it. A two-wavelength, pump-probe transduction scheme is adopted in which a pump laser wavelength-tuned to an optical cavity mode is used to drive the mechanical displacement of the NOMS device through the optical gradient force while a separate probe laser with its wavelength tuned to the maximum sensitivity point (as described in Section 2) is utilized to probe the mechanical motion.[50,51,52] Alternatively, a piezodisk attached to the backside of the sample can also be used

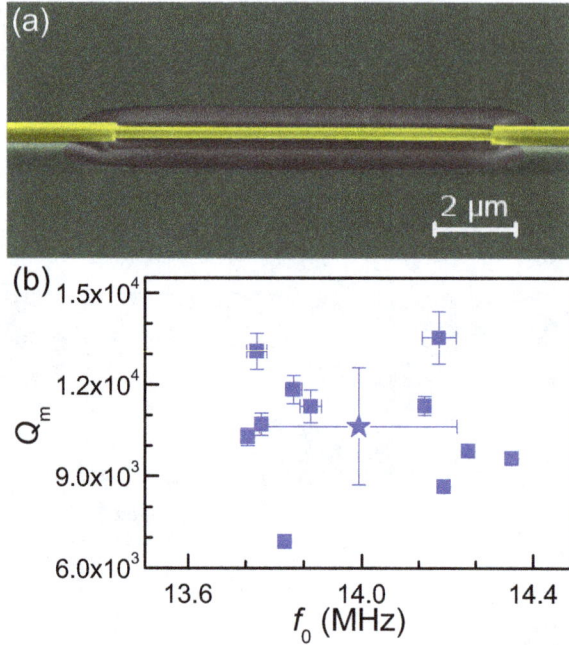

Fig. 7. (a) False-colored SEM image of a NOMS device tested. The device operates by near-field coupled to the partially etched buried oxide layer. (b) Resonance frequencies (f_0) and mechanical quality factors (Q_m) of a series of NOMS doubly clamped beams with the same physical dimensions of $l = 10$ μm, $w = 480$ nm, and $t = 220$ nm. Filled squares represent individual devices whereas the star symbol represents an average of the series. For individual devices, the errors in f_0 and Q_m are fitting errors to Eq. (8). For errors in the averaged behavior, standard deviations in f_0 and Q_m are used. The devices were measured on a free-space interferometry setup, however, the deduced f_0 and Q_m are not affected by the selection of the measurement scheme.

to excite the device.[45] The setup allows fast, and potentially fully-automated probe of a large number of NOMS devices in a single chamber pump-down cycle (provided that nanophotonic circuits connecting various NOMS devices have the same pitch size and the vacuum housing is fitted onto a computer-controlled, two-axis travel stage). Our measurement setup design represents a significant improvement when compared with previous optical-fiber-based measurement systems, in which only a single NOMS device may be tested during one pump-down cycle.[64]

Figure 8 summarizes our measurement results on a representative integrated NOMS device. The device tested consists of a $w = 160$ nm cantilever beam coupled to a race-track resonator. Straight sections in the race-track resonator are 3 μm long and the radius of the arcs is 5 μm.[65] The length of the cantilever is 4.4 μm and it is placed at a distance of 110 nm away from the straight section of the race-track resonator. When the wavelength of the in-coupled light is swept between 1520 and 1600 nm using a tunable diode laser, the intensity of the transmitted light goes through a series of dips corresponding to the race-track optical cavity resonance

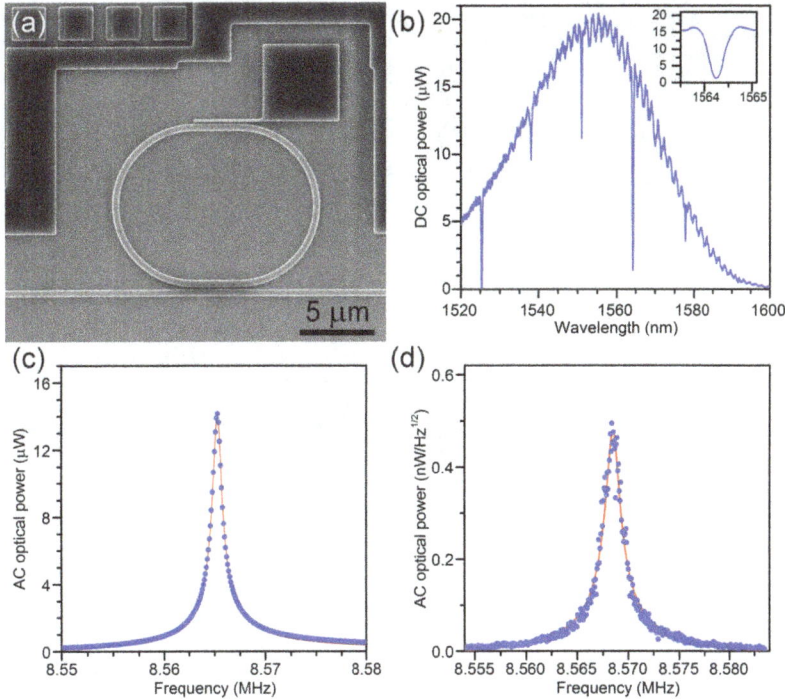

Fig. 8. (a) SEM image of a NOMS device. A 4.4 μm long cantilever is coupled to the straight section of a race-track optical cavity. A bus waveguide is used to couple light into the race-track in the all-pass filter configuration. (b) DC optical transmission spectrum of the device. Inset is a zoomed-in view of the spectrum in the region around 1564 nm. (c) Driven and (d) thermomechanical response of the cantilever measured with NOMS transduction. Filled blue dots are experimental data points whereas red curves are fittings to Eq. (8).

modes, as theoretically described in Section 2. One may compare the experimental results in Fig. 8(b) with theoretical predictions presented in Fig. 1(b). The apparent envelope in the experimental transmission coefficient centered at $\sim$1550 nm is due to the wavelength-dependent coupling efficiency of the grating couplers.[62,68] A zoomed-in view of the cavity resonance mode is presented in the inset of Fig. 8(b) from which a full-width-at-half-maximum of the specific resonance mode of $\sim$0.3 nm can be deduced. Considering a free spectral range of 12.8 nm, the finesse of the race-track resonator is ≈ 43. Figure 8(c) shows the driven response of the cantilever from which a resonance frequency at 8.568 MHz and a mechanical quality factor > 10000 can be inferred. Figure 8(d) is the TM noise response of the same device. Using the calibration procedure described in Section 2, the responsivity, $|\partial T/\partial x|$, is deduced to be 200 nW/nm. This is a rather large value considering the typical noise background of an amplified photodetector for the optical communication C-band being in the range of 3 pW/Hz$^{1/2}$. With a 1 Hz bandwidth, the photodetector-noise-limited sensitivity is $\sim$ 15 fm.

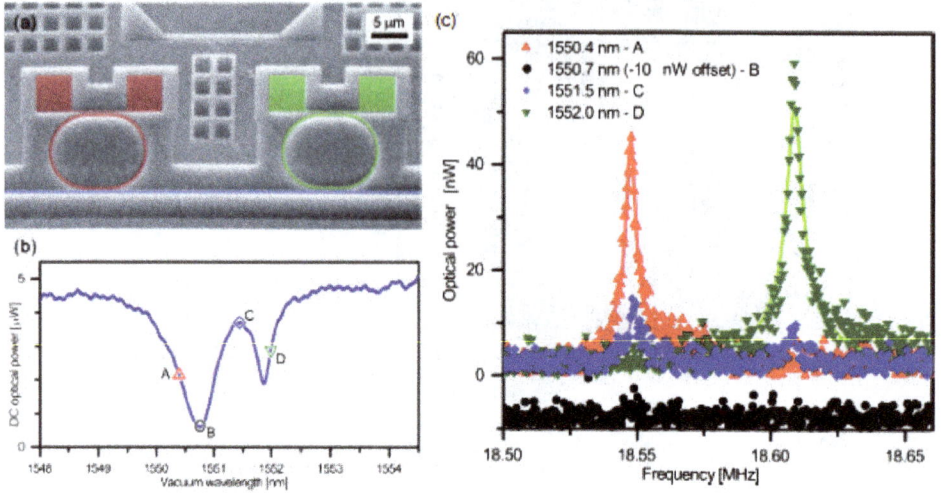

Fig. 9. (a) False-colored SEM image of two NOMS devices coupled to the same bus waveguide, in the wavelength-division multiplexing setting. (b) DC optical transmission spectrum of the two-device multiplexing system as shown in (a). Labels indicate probe laser wavelength used in the experiment. (c) Thermomechanical noise response of the system for different probe wavelengths. Adapted from V. T. K. Sauer, Z. Diao, M. R. Freeman, and W. K. Hiebert, "Wavelength-division multiplexing of nano-optomechanical doubly clamped beam systems", *Opt. Lett.* **40** (2015) 1948–1951 with permission from OSA Publishing.

Many practical applications require not only large-scale integration of loads of NOMS devices onto the same chip but also a well-established and easily-implemented approach to address and readout individual devices. If each device in a large array requires its own connections to function, the photonic circuit design would have been extremely complicated. Luckily, the intrinsic compatibility of NOMS with modern silicon photonics allows them to enjoy all the benefits the later has to offer. One of the most powerful properties of state-of-the-art optical communication systems is their ability to perform wavelength-division multiplexing (WDM). Here, a single waveguide can be used to carry a large number of signal channels and each channel is associated with a predefined color of light, significantly improving the bandwidth of the system. A similar approach can be used to address individual NOMS devices coupled to the same on-chip photonic waveguide. As shown in Fig. 9, two race-track cavities are placed adjacent to a single bus waveguide.[72] Behind each race-track cavity sits a nanomechanical doubly clamped beam. Although both race-track cavities have identical nominal dimensions, the same as those of the device in Fig. 8(a), there are subtle differences in the position of their optical resonance modes and the quality factor, Q_o, due mainly to imperfections in the sample fabrication process.

The DC transmission spectrum of the two-NOMS multiplexing system is shown in Fig. 9(b). Resonance modes of the two race-track cavities partially overlap. When the laser wavelength is positioned at point A, the point with close-to maximum $|\partial T/\partial\lambda|$ for one of the race-track cavities, the transmitted signal is only sensitive to the motion of the NOMS beam coupled to this cavity, as shown in Fig. 9(c). Alternatively, by positioning the laser wavelength at point D, the motion of the other NOMS beam can be readout. For point C, there is sufficient coupling of the input light into both race-track cavities and $|\partial T/\partial\lambda|$ for both cavities is non-zero, thus, both beams can be transduced simultaneously. Conversely, if one positions the laser wavelength at point B, the transmitted light intensity is independent of the motion of either NOMS beams. The reason is that though enough light can be coupled into one of the race-track cavities as in case A, since $|\partial T/\partial\lambda|$ is zero, the transmission coefficient is still insensitive to mechanical motion.

An alternative approach to multiplex many NOMS devices through the same input and output port is to assign signal channels according to the characteristic resonance frequency of individual mechanical devices, the so-called frequency-division

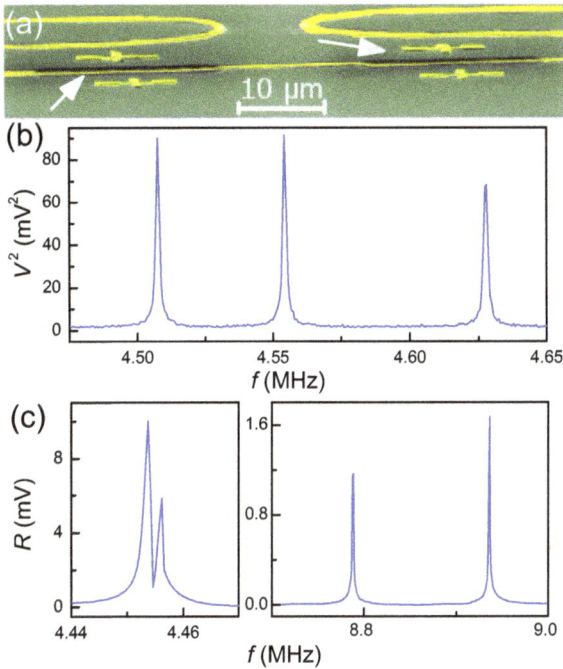

Fig. 10. (a) False-colored SEM image of two doubly clamped beams embedded in the same race-track optical cavity. The two arrows indicate the positions of the beams. (b) TM noise of a three-device multiplexing system. All the doubly clamped beams in the system have the same nominal length of 15 μm. The spectrum was acquired with a spectrum analyzer at a bandwidth of 300 Hz. (c) Driven response of a four-device multiplexing system. The two longer beams are 15 μm long by design whereas the two shorter beams are $\approx$ 10 μm in length. The device was actuated by a piezodisk driven with 100 mV$_{\text{p-p}}$. The spectrum was recorded using a high-frequency lock-in amplifier (Zurich Instruments HF2LI) with an equivalent noise bandwidth of 7.7 Hz and 4 averages.

multiplexing. Such a readout scheme has been widely adopted in the electronics domain to connect sensors in a large matrix, *e.g.* in the design of superconducting microwave kinetic inductance detectors for x-ray imaging.[73] To enhance the transduction responsivity, several NOMS beams of different resonance frequencies may be embedded into the same race-track optical resonator, as shown in Fig. 10(a). Here, two separate sections in the same race-track cavity have been released to form two doubly clamped beams. When they move towards and away from the partially etched buried oxide layer beneath, their effective refractive indices change which in turn causes a shift of the optical cavity mode. Since the governing mechanism in NOMS displacement readout is phase interaction rather than light scattering,[74] introducing multiple devices into the same race-track optical cavity will not noticeably decrease the cavity finesse and the transduction efficiency. Figure 10(b) shows the thermomechanical noise spectrum of a three-device multiplexing system similar to that shown in Fig. 10(a). TM noise modes of all three beams can be observed with high signal-to-noise ratio. Figure 10(c) displays driven responses of another device where four NOMS beams can be multiplexed. The resonance frequencies of the two longer beams ($l = 15$ μm) are very close and they partially overlap in the spectrum whereas resonance modes of the two shorter beams with $l \approx 10$ μm are sufficiently different to be completely separated.

A suitable multiplexing scheme should be chosen according to the application in concern. Wavelength-division multiplexing allows one to address individual NOMS devices independent of their respective mechanical resonance frequencies. This is especially useful when the device mechanical resonance frequency itself carries the signal of interest, *e.g.* in case of inertial mass sensing[75,76,77,78,79] and frequency-tracking thermometry and magnetometery.[80,81,82] On the other hand, mechanical frequency-division multiplexing has the potential to drastically reduce the footprint of the device matrix, and is suitable for applications such as clocking and data storage where the device mechanical resonance frequency does not vary during operation.

5. Conclusions

Nanomechanical resonators are versatile devices which have found their applications in a large variety of domains, among which inertial mass sensing has generated massive ongoing interests. Operating in the frequency-tracking mode, absorbed mass induces a shift in the resonance frequency of nanomechanical devices, and the mass sensitivity scales proportionally with the effective mass of the nanomechanical resonator.[83]

$$\delta m \approx -2\frac{m_{\text{eff}}}{f_0}\delta f. \tag{10}$$

Constructing ever more sensitive nanomechanical mass sensors requires smaller and smaller resonators operating at higher and higher frequencies. The extremely high responsivity and ultra-wide bandwidth offered by NOMS displacement transduction are in dear demand against this backdrop. NOMS fabrication can be conducted at state-of-the-art silicon photonics foundries, providing a clear path way to chip-level integration. The device multiplexing schemes discussed in this work, namely, wavelength-division multiplexing and frequency-division multiplexing, can further reduce the design complexity and increase the system capacity, both crucial parameters to be considered in emerging large-scale integrated photonic circuits.

An existing obstacle which has so-far delayed the wide-scale penetration of silicon photonics as well as NOMS into commercial products is the lack of reliable and low-cost on-chip light sources and detectors. Recently, encouraging developments have been made on both fronts, represented by the realization of on-chip electrically-pumped germanium lasers[84] and epitaxial growth of III-V semiconductors on SOI substrates with potential applications in photodetectors.[85] It can be envisaged that such developments will lead to highly-intelligent integrated silicon photonic systems that are capable of generating, processing, transmitting, and storing information solely with light in the near future. They can also pave the way towards monolithic integration of silicon photonics with micro-/nanoelectronics. Further developments of NOMS will no doubt benefit tremendously from this trend. They are to become an indispensable component of next-generation silicon photonic chips, as sensors, clock references, solid-state switches, and memory units, *etc.*

Acknowledgments

We would like to acknowledge the National Institute for Nanotechnology, Alberta Innovates Technology Futures, Alberta Innovates Health Solutions, the Natural Sciences and Engineering Research Council, Canada, the Canadian Institute for Advanced Research, and CMC Microsystems which funded this work. The fabrication of the devices was facilitated through CMC Microsystems, and post processing was conducted at the University of Alberta nanoFAB. Device imaging was performed on microscopy facilities located at the University of Alberta NanoFab and the National Institute for Nanotechnology. We are indebted to Professor Mark R. Freeman at the University of Alberta, Dr. Lukas Chrostowski and Professor Nicolas A. F. Jaeger at the University of British Columbia, and Dan Deptuck and Jessica Zhang at CMC Microsystems for fruitful discussions. Z. D. would like to thank the support he received from the Department of Mathematics, Physics and Electrical Engineering of Halmstad University, Sweden during the preparation of this manuscript.

References

1. M. M. Waldrop, "The chips are down for Moore's law", *Nature*, **530** (2016) 144–147.
2. H. G. Craighead, "Nanoelectromechanical systems", *Science* **290** (2000) 1532–1535.
3. K. L. Ekinci and M. L. Roukes, "Nanoelectromechanical systems", *Rev. Sci. Instrum.* **76** (2005) 061101.
4. K. Srinivasan, H. Miao, M. T. Rakher, M. Davanço, and V. Aksyuk, "Optomechanical transduction of an integrated silicon cantilever probe using a microdisk resonator", *Nano Lett.* **11** (2011) 791–797.

5. Y. Liu, H. Miao, V. Aksyuk, and K. Srinivasan, "Wide cantilever stiffness range cavity optomechanical sensors for atomic force microscopy", *Opt. Express* **20** (2012) 18268–18280.

6. A. K. Naik, M. S. Hanay, W. K. Hiebert, X. L. Feng, and M. L. Roukes, "Towards single-molecule nanomechanical mass spectrometry", *Nat. Nanotech.* **4** (2009) 445–450.

7. V. T. K. Sauer, M. R. Freeman, and W. K. Hiebert, "Device overshield for mass-sensing enhancement (DOME) structure fabrication", *J. Micromech. Microeng.* **20** (2010) 105020.

8. J. Chaste, A. Eichler, J. Moser, G. Ceballos, R. Rurali, and A. Bachtold, "A nanomechanical mass sensor with yoctogram resolution", *Nat. Nanotech.* **7** (2012) 301–304.

9. M. S. Hanay, S. I. Kelber, C. D. O'Connell, P. Mulvaney, J. E. Sader, and M. L. Roukes "Inertial imaging with nanomechanical systems", *Nat. Nanotech.* **10** (2015) 339–344.

10. J. Losby, J. A. J. Burgess, Z. Diao, D. C. Fortin, W. K. Hiebert, and M. R. Freeman, "Thermo-mechanical sensitivity calibration of nanotorsional magnetometers", *J. Appl. Phys.* **111** (2012) 07D305.

11. S. Forstner, S. Prams, J. Knittel, E. D. van Ooijen, J. D. Swaim, G. I. Harris, A. Szorkovszky, W. P. Bowen, and H. Rubinsztein-Dunlop, "Cavity optomechanical magnetometer", *Phys. Rev. Lett.* **108** (2012) 120801.

12. J. A. J. Burgess, A. E. Fraser, F. Fani Sani, D. Vick, B. D. Hauer, J. P. Davis, and M. R. Freeman, "Quantitative magneto-mechanical detection and control of the Barkhausen effect", *Science* **339** (2013) 1051–1054.

13. Z. Diao, J. E. Losby, J. A. J. Burgess, V. T. K. Sauer, W. K. Hiebert, and M. R. Freeman, "Stiction-free fabrication of lithographic nanostructures on resist-supported nanomechanical resonators", *J. Vac. Sci. Technol. B* **31** (2013) 051805.

14. J. E. Losby, F. Fani Sani, D. T. Grandmont, Z. Diao, M. Belov, J. A. J. Burgess, S. R. Compton, W. K. Hiebert, D. Vick, K. Mohammad, E. Salimi, G. E. Bridges, D. J. Thomson, and M. R. Freeman, "Torque-mixing magnetic resonance spectroscopy", *Science* **350** (2015) 798–801.

15. X. C. Zhang, E. B. Myers, J. E. Sader, and M. L. Roukes, "Nanomechanical torsional resonators for frequency-shift infrared thermal sensing", *Nano Lett.* **13** (2013) 1528–1534.

16. S. Schmid, K. Wu, P. E. Larsen, T. Rindzevicius, and A. Boisen, "Low-power photothermal probing of single plasmonic nanostructures with nanomechanical string resonators", *Nano Lett.* **14** (2014) 2318–2321.

17. M. Bagheri, M. Poot, M. Li, W. P. H. Pernice, and H. X. Tang, "Dynamic manipulation of nanomechanical resonators in the high-amplitude regime and non-volatile mechanical memory operation", *Nat. Nanotech.* **6** (2011) 726–732.

18. A. G. Krause, M. Winger, T. D. Blasius, Q. Lin, and O. Painter, "A high-resolution microchip optomechanical accelerometer", *Nat. Photon.* **6** (2012) 768–772.

19. D. Antonio, D. H. Zanette, and D. López, "Frequency stabilization in nonlinear micromechanical oscillators", *Nat. Commun.* **3** (2011) 806.

20. P. A. Truitt, J. B. Hertzberg, C. C. Huang, K. L. Ekinci, and K. C. Schwab, "Efficient and sensitive capacitive readout of nanomechanical resonator arrays", *Nano Lett.* **7** (2007) 120–126.

21. J. Sulkko, M. A. Sillanpää, P. Häkkinen, L. Lechner, M. Helle, A. Fefferman, J. Parpia, and P. J. Hakonen, "Strong gate coupling of high-Q nanomechanical resonators", *Nano Lett.* **10** (2010) 4884–4889.

22. J. R. Montague, K. A. Bertness, N. A. Sanford, V. M. Bright, and C. T. Rogers, "Temperature-dependent mechanical-resonance frequencies and damping in ensembles of gallium nitride nanowires", *Appl. Phys. Lett.* **101** (2012) 173101.

23. R. G. Beck, M. A. Eriksson, R. M. Westervelt, K. L. Campman, and A. C. Gossard, "Strain-sensing cryogenic field-effect transistor for integrated strain detection in GaAs/AlGaAs microelectromechanical systems", *Appl. Phys. Lett.* **68** (1996) 3763–3765.

24. R. B. Karabalin, M. H. Matheny, X. L. Feng, E. Defaÿ, G. Le Rhun, C. Marcoux, S. Hentz, P. Andreucci, and M. L. Roukes, "Piezoelectric nanoelectromechanical resonators based on aluminum nitride thin films", *Appl. Phys. Lett.* **95** (2009) 103111.

25. M. Li, H. X. Tang, and M. L. Roukes, "Ultra-sensitive NEMS-based cantilevers for sensing, scanned probe and very high-frequency applications", *Nat. Nanotech.* **2** (2007) 114–120.

26. R. He, X. L. Feng, M. L. Roukes, and P. Yang, "Self-transducing silicon nanowire electromechanical systems at room temperature", *Nano Lett.* **8** (2008) 1756–1761.

27. A. Koumela, S. Hentz, D. Mercier, C. Dupré, E. Ollier, P. X.-L. Feng, S. T. Purcell, and L. Duraffourg, "High frequency top-down junction-less silicon nanowire resonators", *Nanotechnology* **24** (2013) 435203.

28. U. Kemiktarak, T. Ndukum, K. C. Schwab, and K. L. Ekinci, "Radio-frequency scanning tunnelling microscopy", *Nature* **450** (2007) 85–88.

29. N. E. Flowers-Jacobs, D. R. Schmidt, and K. W. Lehnert, "Intrinsic noise properties of atomic point contact displacement detectors", *Phys. Rev. Lett.* **98** (2007) 096804.

30. I. Bargatin, I. Kozinsky, and M. L. Roukes, "Efficient electrothermal actuation of multiple modes of high-frequency nanoelectromechanical resonators", *Appl. Phys. Lett.* **90** (2007) 093116.

31. R. G. Knobel and A. N. Cleland, "Nanometre-scale displacement sensing using a single electron transistor", *Nature* **424** (2003) 291–293.

32. M. D. LaHaye, O. Buu, B. Camarota, and K. C. Schwab, "Approaching the quantum limit of a nanomechanical resonator", *Science* **304** (2004) 74–77.

33. Y. Oda, K. Onomitsu, R. Kometani, S. Warisawa, S. Ishihara, and H. Yamaguchi, "Electromechanical displacement detection with an on-chip high electron mobility transistor amplifier", *Jpn. J. Appl. Phys.* **50** (2011) 06GJ01.

34. M. Faucher, Y. Cordier, M. Werquin, L. Buchaillot, C. Gaquière, and D. Théron, "Electromechanical transconductance properties of a GaN MEMS resonator with fully integrated HEMT transducers", *J. Microelectromech. Syst.* **21** (2012) 370–378.

35. B. Ilic, S. Krylov, K. Aubin, R. Reichenbach, and H. G. Craighead, "Optical excitation of nanoelectromechanical oscillators", *Appl. Phys. Lett.* **86** (2005) 193114.

36. S. S. Verbridge, D. F. Shapiro, H. G. Craighead, and J. M. Parpia, "Macroscopic tuning of nanomechanics: substrate bending for reversible control of frequency and quality factor of nanostring resonators", *Nano Lett.* **7** (2007) 1728–1735.

37. W. K. Hiebert, D. Vick, V. Sauer, and M. R. Freeman, "Optical interferometric displacement calibration and thermomechanical noise detection in bulk focused ion beam-fabricated nanoelectromechanical systems", *J. Micromech. Microeng.* **20** (2010) 115038.

38. A. Sampathkumar, K. L. Ekinci, and T. W. Murray, "Multiplexed optical operation of distributed nanoelectromechanical systems arrays", *Nano Lett.* **11** (2011) 1014–1019.

39. R. van Leeuwen, D. M. Karabacak, H. S. J. van der Zant, and W. J. Venstra, "Nonlinear dynamics of a microelectromechanical oscillator with delayed feedback", *Phys. Rev. B* **88** (2013) 214301.

40. T. Kouh, D. Karabacak, D. H. Kim, and K. L. Ekinci, "Diffraction effects in optical interferometric displacement detection in nanoelectromechanical systems", *Appl. Phys. Lett.* **86** (2005) 013106.

41. I. De Vlaminck, J. Roels, D. Taillaert, D. Van Thourhout, R. Baets, L. Lagae, and G. Borghs, "Detection of nanomechanical motion by evanescent light wave coupling", *Appl. Phys. Lett.* **90** (2007) 233116.

42. G. Anetsberger, O. Arcizet, Q. P. Unterreithmeier, R. Rivière, A. Schliesser, E. M. Weig, J. P. Kotthaus, and T. J. Kippenberg, "Near-field cavity optomechanics with nanomechanical oscillators", *Nat. Phys.* **5** (2009) 909–914.

43. M. Li, W. H. P. Pernice, and H. X. Tang, "Reactive cavity optical force on microdisk-coupled nanomechanical beam waveguides", *Phys. Rev. Lett.* **103** (2009) 223901.

44. W. H. P. Pernice, C. Xiong, C. Schuck, and H. X. Tang, "High-Q aluminum nitride photonic crystal nanobeam cavities", *Appl. Phys. Lett.* **100** (2012) 091105.

45. V. T. K. Sauer, Z. Diao, M. R. Freeman, and W. K. Hiebert, "Nanophotonic detection of side-coupled nanomechanical cantilevers", *Appl. Phys. Lett.* **100** (2012) 261102.

46. A. Schliesser, G. Anetsberger, R. Rivière, O. Arcizet, and T. J. Kippenberg, "High-sensitivity monitoring of micromechanical vibration using optical whispering gallery mode resonators", *New J. Phys.* **10** (2008) 095015.

47. M. Eichenfield, R. Camacho, J. Chan, K. J. Vahala, and O. Painter, "A picogram- and nanometre-scale photonic-crystal optomechanical cavity", *Nature* **459** (2009) 550–555.

48. M. Eichenfield, J. Chan, R. M. Camacho, K. J. Vahala, and O. Painter, "Optomechanical crystals", *Nature* **462** (2009) 78–82.

49. J. Chan, T. P. M. Alegre, A. H. Safavi-Naeini, J. T. Hill, A. Krause, S. Gröblacher, M. Aspelmeyer, and O. Painter, "Laser cooling of a nanomechanical oscillator into its quantum ground state", *Nature* **478** (2011) 89–92.

50. M. Li, W. H. P. Pernice, C. Xiong, T. Baehr-Jones, M. Hochberg, and H. X. Tang, "Harnessing optical forces in integrated photonic circuits", *Nature* **456** (2008) 480–484.

51. M. Li, W. H. P. Pernice, and H. X. Tang, "Tunable bipolar optical interactions between guided lightwaves", *Nat. Photon.* **3** (2009) 464–468.

52. J. Roels, I. De Vlaminck, L. Lagae, B. Maes, D. Van Thourhout, and R. Baets, "Tunable optical forces between nanophotonic waveguides", *Nat. Nanotech.* **4** (2009) 510–513.

53. D. Van Thourhout and J. Roels, "Optomechanical device actuation through the optical gradient force", *Nat. Photon.* **4** (2010) 211–217.

54. W. Bogaerts, M. Fiers, and P. Dumon, "Design challenges in silicon photonics", *IEEE J. Sel. Topics Quantum Electron.* **20** (2014) 8202008.

55. S. Gigan, H. R. Böhm, M. Paternostro, F. Blaser, G. Langer, J. B. Hertzberg, K. C. Schwab, D. Bäuerle, M. Aspelmeyer, and A. Zeilinger, "Self-cooling of a micromirror by radiation pressure", *Nature* **444** (2006) 67–70.

56. O. Arcizet, P.-F. Cohadon, T. Briant, M. Pinard, and A. Heidmann, "Radiation-pressure cooling and optomechanical instability of a micromirror", *Nature* **444** (2006) 71–74.

57. T. J. Kippenberg and K. J. Vahala, "Cavity opto-mechanics", *Opt. Express* **15** (2007) 17172–17205.

58. A. D. O'Connell, M. Hofheinz, M. Ansmann, R. C. Bialczak, M. Lenander, E. Lucero, M. Neeley, D. Sank, H. Wang, M. Weides, J. Wenner, J. M. Martinis, and A. N. Cleland, "Quantum ground state and single-phonon control of a mechanical resonator", *Nature* **464** (2010) 697–703.

59. M. Aspelmeyer, T. J. Kippenberg, and F. Marquardt, "Cavity optomechanics", *Rev. Mod. Phys.* **86** (2014) 1391–1452.

60. M. Aspelmeyer, T. J. Kippenberg, and F. Marquardt eds., *Cavity Optomechanics: Nano- and Micromechanical Resonators Interacting with Light*, Springer, Berlin and Heidelberg, 2014.

61. M. Li, W. H. P. Pernice, and H. X. Tang, "Broadband all-photonic transduction of nanocantilevers", *Nat. Nanotech.* **4** (2009) 377–382.

62. G. Roelkens, D. Vermeulen, D. Van Thourhout, R. Baets, S. Brision, P. Lyan, P. Gautier, and J.-M. Fédéli, "High efficiency diffractive grating couplers for interfacing a single mode optical fiber with a nanophotonic silicon-on-insulator waveguide circuit", *Appl. Phys. Lett.* **92** (2008) 131101.

63. P. Dumon, "Ultra-compact integrated optical filters in silicon-on-insulator by means of wafer-scale technology", PhD thesis, Ghent University, 2007.

64. J. Roels, "Actuation of integrated nanophotonic devices through the optical gradient force", Ph.D. thesis, Ghent University, 2011.

65. V. T. K. Sauer, Z. Diao, M. R. Freeman, and W. K. Hiebert, "Optical racetrack resonator transduction of nanomechanical cantilevers", *Nanotechnology* **25** (2014) 055202.

66. E. Hecht, *Optics* (4th ed.), Addison-Wesley, San Francisco, CA, 2002, pp. 420–423.

67. A. N. Cleland, *Foundations of Nanomechanics: From Solid-State Theory to Device Applications*, Springer, Berlin and Heidelberg, 2003.

68. Z. Diao, J. E. Losby, V. T. K. Sauer, J. N. Westwood, M. R. Freeman, and W. K. Hiebert, "Confocal scanner for highly sensitive photonic transduction of nanomechanical resonators", *Appl. Phys. Express* **6** (2013) 065202.

69. M. Hossein-Zadeh and K. J. Vahala, "Observation of optical spring effect in a microtoroidal optomechanical resonator", *Opt. Lett.* **32** (2007) 1611–1613.

70. G. D. Cole, I. Wilson-Rae, K. Werbach, M. R. Vanner, and M. Aspelmeyer, "Phonon-tunnelling dissipation in mechanical resonators", *Nat. Commun.* **2** (2011) 231.

71. M. Imboden and P. Mohanty, "Dissipation in nanoelectromechanical systems", *Phys. Rep.* **534** (2014) 89–146.

72. V. T. K. Sauer, Z. Diao, M. R. Freeman, and W. K. Hiebert, "Wavelength-division multiplexing of nano-optomechanical doubly clamped beam systems", *Opt. Lett.* **40** (2015) 1948–1951.

73. G. Ulbricht, B. A. Mazin, P. Szypryt, A. B. Walter, C. Bockstiegel, and B. Bumble, "Highly multiplexible thermal kinetic inductance detectors for x-ray imaging spectroscopy", *Appl. Phys. Lett.* **106** (2015) 251103.

74. O. Basarir, S. Bramhavar, and K. L. Ekinci, "Motion transduction in nanoelectromechanical (NEMS) arrays using near-field optomechanical coupling", *Nano. Lett.* **12** (2012) 534–539.

75. S. Dohn, S. Schmid, F. Amiot, and A. Boisen, "Position and mass determination of multiple particles using cantilever based mass sensors", *Appl. Phys. Lett.* **97** (2010) 044103.

76. E. Gil-Santos, D. Ramos, J. Martínez, M. Fernández-Regúlez, R. García, Á. San Paulo, M. Calleja, and J. Tamayo, "Nanomechanical mass sensing and stiffness spectrometry based on two-dimensional vibrations of resonant nanowires", *Nat. Nanotech.* **5** (2010) 641–645.

77. M. S. Hanay, S. Kelber, A. K. Naik, D. Chi, S. Hentz, E. C. Bullard, E. Colinet, L. Duraffourg, and M. L. Roukes, "Single-protein nanomechanical mass spectrometry in real time", *Nat. Nanotech.* **7** (2012) 602–608.

78. E. Sage, A. Brenac, T. Alava, R. Morel, C. Dupré, M. S. Hanay, M. L. Roukes, L. Duraffourg, C. Masselon, and S. Hentz, "Neutral particle mass spectrometry with nanomechanical systems", *Nat. Commun.* **6** (2015) 6482.

79. S. Olcum, N. Cermak, S. C. Wasserman, and S. R. Manalis, "High-speed multiple-mode mass-sensing resolves dynamic nanoscale mass distributions", *Nat. Commun.* **6** (2015) 7070.

80. B. C. Stipe, H. J. Mamin, T. D. Stowe, T. W. Kenny, and D. Rugar, "Magnetic dissipation and fluctuations in individual nanomagnets measured by ultrasensitive cantilever magnetometry", *Phys. Rev. Lett.* **86** (2001) 2874–2877.

81. D. P. Weber, D. Rüffer, A. Buchter, F. Xue, E. Russo-Averchi, R. Huber, P. Berberich, J. Arbiol, A. Fontcuberta i Morral, D. Grundler, and M. Poggio, "Cantilever magnetometry of individual Ni nanotubes", *Nano Lett.* **12** (2012) 6139–6144.

82. A. Mehlin, F. Xue, D. Liang, H. F. Du, M. J. Stolt, S. Jin, M. L. Tian, and M. Poggio, "Stabilized skyrmion phase detected in MnSi nanowires by dynamic cantilever magnetometry", *Nano Lett.* **15** (2015) 4839–4844.

83. K. L. Ekinci, Y. T. Yang, and M. L. Roukes, "Ultimate limits to inertial mass sensing based upon nanoelectromechanical systems", *J. Appl. Phys.* **95** (2004) 2682–2689.

84. R. E. Camacho-Aguilera, Y. Cai, N. Patel J. T. Bessette, M. Romagnoli, L. C. Kimerling, and J. Michel, "An electrically pumped germanium laser", *Opt. Express* **20** (2012) 11316–11320.

85. H. Schmid, M. Borg, K. Moselund, L. Gignac, C. M. Breslin, J. Bruley, D. Cutaia, and H. Riel, "Template-assisted selective epitaxy of III-V nanoscale devices for co-planar heterogeneous integration with Si", *Appl. Phys. Lett.* **106** (2015) 233101.

Author Index

www.ingramcontent.com/pod-product-compliance
Lightning Source LLC
Chambersburg PA
CBHW081519190326
41458CB00015B/5405